Synthesis Lectures on Ocean Systems Engineering

Series Editor

Nikolas Xiros, University of New Orleans, New Orleans, LA, USA

The series publishes short books on state-of-the-art research and applications in related and interdependent areas of design, construction, maintenance and operation of marine vessels and structures as well as ocean and oceanic engineering.

Gerd Würsig

The Safety Principles for the Use of Low Flashpoint Fuels in Shipping

 Springer

Gerd Würsig
GMW Consultancy
Hammah, Niedersachsen, Germany

ISSN 2692-4420   ISSN 2692-4471 (electronic)
Synthesis Lectures on Ocean Systems Engineering
ISBN 978-3-031-64173-2   ISBN 978-3-031-64174-9 (eBook)
https://doi.org/10.1007/978-3-031-64174-9

This Springer imprint is published by the registered company Springer Nature Switzerland AG
The registered company address is: Gewerbestrasse 11, 6330 Cham, Switzerland

If disposing of this product, please recycle the paper.

The author's hope is that this publication may support the daily work of all those who are designing, building and operating gas carriers and ships which use fuel alternatives to the well known fuel oil.

Preface

The "International Code of Safety for Ships using Gases or other Low-Flashpoint Fuels"[1] [1] is the International Maritime Organization's (IMO) instrument to regulate the safety of handling fuels which were not commonly used in shipping prior to the introduction of this Code in 2015. The more precise date of the milestone for the so called "IGF-Code" is the afternoon on Friday 12th of June 2015 when the IMO MSC-95 meeting ended and the Adoption of the IGF-Code became a reality. It may be a coincidence but for the author, the fact that Meyer Werft and Carnival Corporation launched a press release about the building of four LNG fuelled Cruise ships after the weekend on Monday 15th of June 2015, highlights the relevance of the IGF-Code to shipping (comp. Annex, p.125).

At the beginning of the development in 2004 the Norwegian Administration proposed a Guideline for Gas as fuel assuming that Methane stored as LNG would be the only gas used as fuel for ships like on the first Norwegian ferry operated with LNG as fuel the "MS GLUTRA" (Fig. 1).[2]

Fig. 1 First Norwegian non-LNG Carrier fuelled with LNG as fuel: MS GLUTRA September 2003. (*Source* Dr. Gerd Wuersig)

---

[1] *Flashpoint* comp. Glossary, p. 129.
[2] The Norwegian proposal to the IMO was based on the experience with the "MS Glutra".

Consequently, the IMO working group installed to develop an International Code named it "International Gas as Fuel-Code" or "IGF-Code". This acronym became very famous. No one liked to change it even though it became clear that other unconventional fuels for shipping need new regulations. For this very human, practical reason the acronym "IGF-Code" was saved by renaming it the "International Code of Safety for Ships using Gases or other Low-Flashpoint Fuels". It is a little bit inconsistent that the word "Code" is the second word of the title but is at the end of the acronym.[3]

The anecdote about the naming of the IGF-Code illustrates a motivation for writing this publication. Technical regulations like the IGF-Code [1] are developed by large groups of experts over long periods of time. The explanation of the background for the requirements of such rules is not part of the development process. Consequently, very often the reasoning behind the rules is lost over time or at least hidden from the end user.

This publication aims to explain the safety principles behind the rules for ships which use unconventional fuels.[4] The hope of the author is that the understanding of the safety principles may contribute to the technological based interpretation of the rules and their further development.

The reader should note that this publication is written to explain the subject and not as a scientific textbook. Some parts may be seen by some "professionals" to be "too simple", "incomplete", "not meeting the scientific standards" but this is because the aim is not to write a scientific work. Even if the subject may be seen as a "very dry" one, the author wanted to have some fun writing it down. This may be one explanation.[5]

Those who might be interested in the development of the basic IGF-Code may read the keynote from Motorship Conference—Gas fuelled ships—in Hamburg from 10th to 12th November 2015 [2].

The author assumes no responsibility or liability for any errors or omissions in the content of the book. Neither can he guarantee completeness, accuracy, usefulness or timeliness. In particular, the author is not responsible for any errors or omissions, or for the results obtained from the use of this information.

Hammah, Germany   Dr.-Ing. Gerd Würsig[6]
March 2023   Dr.Ing.Wuersig@ewe.net

---

[3] Note that IMO CCC-8 in 2022 decided to change the wording from "Low-Flashpoint Fuels" to "Alternative Fuels". Time will show what will happen to the title of the IGF-Code.

[4] In this context "conventional fuels" are the oil based fuels used prior to LNG and the other "low flashpoint"/"alternative fuels".

[5] Life is too seriously to take it serious (Google "thinks" that Oscar Wilde may have written this).

[6] My name is written "Würsig" in German. But this looks to me to be too unconventional for most readers.

References

1. IMO (2016), IMO Resolution MSC.391(95), IGF-Code: International Code of Safety for Ships using Gases or other Low-Flashpoint Fuels, IMO, London, ISBN 978-92-801-1653-3
2. Wuersig, Gerd (2015); The development of THE INTERNATIONAL CODE OF SAFETY FOR SHIPS USING GASES OR OTHER LOW-FLASHPOINT FUELS (IGF CODE)—IMO MSC.391(95); Motorship Conference—Gas fuelled ships—Hamburg, 10/12 Nov. 2015

References

Acknowledgements

First of all my thanks go to my wife Regina. She had to be very patient with me when I was spending our weekends for writing this book.

For sharing their technical knowledge and deep understanding of safety for gas handling and storage systems, I thank the colleagues I had the possibility to work with at Germanischer Lloyd, German Shipping Administration, IMO, SIGTTO and SGMF. My special thank is given to my former colleague at Germanischer Lloyd Martin Böckenhauer (†, 2006) who introduced me into the subject of gas carrier safety and IMO work and my longtime line manager Dr. Reinhard Krapp who gave me the possibility to work on these subjects.

For supporting this publication with illustrations I thank all involved individuals and companies. For the constructive discussions and technical hints related to this book, I like to thank all who gave me the opportunity to ask them for their views. Specially I thank Dr. Hans-Christian Haarmann-Kühn, Head of Engineering at TGE Marine.

For proofreading I like to thank Michael Wright, excelsior p.s., for his hard work. The parts which still sound German are my responsibility and definitely not related to his work.

Contents

Acronyms

API	American Petroleum Institute (USA Institution developing saftey standards in refinery industry)
BAM	Bundesanstalt für Materialforschung und –prüfung
BOG	Boil Off Gas
BOR	Boil Off Rate in $\%/d$ at 100% filling
CC	Carbon Capture
CCU	Carbon Capture and Use
CGA	Compressed Gas Association (USA Institution developing saftey standards for transport and use of flammable gases)
CH_3OH, C_2H_5OH	Methaol, Ethanol
CH_4	Methane
CNG	Compressed Natural Gas
CO_2	Carbon Dioxide
DEKRA	Deutscher Kraftfahrzeug-Überwachungs-Verein
DNV	Det Norske Veritas (Classification Society based in Norway)
Double-Block and Bleed	Is a valve arrangement with two block and one bleed valve
DQRDC	Dry Quick Release/Disconnect Coupling
ESD	Emergency Shut Down
ESD-Concept	Emergency Shut Down engine room concept according IGF-Code (2016), Sect. 5.6
ESD-ER	ESD engine room
EX-Z	Explosion Zone
FL	Tank Filling Limit at maximum permitted pressure
FMEA	Failure Mode and Effect Analysis
FR	Functional Requirement
FRs	Functional Requirements
FT	Fischer Tropsch
Gas-Save-ER	Gas safe or inherent safe engine room

GHG	Green House Gas
GVU	Gas Valve Unit
HAZID	HAZard IDentification study
HFO	Heavy Fuel Oil
IATA	The International Air Transport Association
IEC	INTERNATIONAL ECTROTECHNICAL COMMISSION
IEC-60079-10-1	Explosive gas atmospheres; IEC Standard 60079-10-1, 2021
IGC-Code	International Code for the Construction and Equipment of Ships Carrying Liquefied Gases in Bulk
IGF-Code	International Code of Safety for Ships using Gases or other Low-Flashpoint Fuels
IGF-IG	IMO interim guidelines on gas as ship fuel
IMDG-Code	IMO International Maritime Dangerous Goods Code
LEL	Lower Explosion Limit
LFF	Low Flashpoint Fuel
LH$_2$	Liquefied Hydrogen
LL	Tank Loading Limit
LNG	Liquefied Natural Gas
LNH$_3$	Liquefied Ammonia
LPG	Liquefied Petroleum Gas (mixture of Propane unad Butane)
MARVS	Maximum Allowable Relief Valve Setting
MAWP	Maximum Allowable Working Pressure
MDO	Marine Diesel Oil (mixture of mainly destilate fuel with heavy oil)
MGO	Marine Gas Oil (destilate fuel)
MSC	IMO Marine Safety Committee
NH$_3$	Ammonia
NO$_x$	Nitrogen Oxide
Part-A1	IGF-Code (2016) Part-A1 "Specific requirements for ships using natural gas as fuel"
PM	Particulate Matter
PRV	Pressure Relief Valve
PRVs	Pressure Relief Valves
PtX	Power to X
PtX-FT	Power to X Fischer Tropsch fuel
PtX-LMG	Power to X Liquefied Methane Gas
QRA	Quantitative Risk Assessment
QRR	Qualitative Risk Ranking
SIGTTO	Society of Gas Tankers and Terminal Operators
SO$_2$	Sulphure Dioxide
SOLAS	Safety Of Life At Sea convention

S-QRA	Semi-Quantitative Risk Assessment
TÜV	Technischer Überwachungsverein
tank connection space	The tank connection space includes all tank connections of the fuell tank including the first valve outside of the tank
UN	United Nations
WG	Working Groups are set up by IMO committees to work on a subject during IMO meetings

List of Figures

List of Tables

Introduction to Relevant Rules, Guidelines and Fuels

IMO regulations are often written as prescriptive rules with some elements of goal-based rules. A short basic description of the difference between prescriptive and goal-based requirements and the application of these principles in the IGF-Code is given in the beginning of this section. The general applicability of Part A and the special applicability of Part A1 of the IGF-Code is also discussed.[1]

Based on experiences with conventional oil based fuels, the background for Methane as ship fuel and other alternative fuels which currently introduced is highlighted.

1.1 What are Prescriptive and Goal Based Requirements?

The terms "prescriptive", "goal based" named in the heading of this section are often regarded as ultimate and mutually exclusive definitions for the "real" type of rules and regulations. For this reason one expert opinion is that rules either can be prescriptive or goal based.

The author's experience is that, also with regard to the different schools how to write and to apply rules, compromise is the best approach. Also, in this case "the truth is found in the middle".

To be able to use the terms prescriptive and goal based in relation to the requirements in rules and standards the understanding for the use in this publication is given in the following.

Prescriptive requirements give instructions how a subject has to be designed, handled to fulfil the aim which nearly always is behind the requirement. The explanation of the aim for a requirement is in practice never part of the requirement. For this reason, the real sense of the requirement often get lost over time.

---

[1] For the IGF-Code structure see Sect. 1.2, p. 3.

© The Author(s), under exclusive license to Springer Nature Switzerland AG 2025   1
G. Würsig, *The Safety Principles for the Use of Low Flashpoint Fuels in Shipping*,
Synthesis Lectures on Ocean Systems Engineering,
https://doi.org/10.1007/978-3-031-64174-9_1

The very useful regulations for the Pressure Relief Valve sizing (PRV) in the IGF-Code and the IGC-Code [1, 2] as discussed in Sect. 7, are a perfect example of this kind of rule.

Very often these kinds of rules are based on a very limited number of existing cases. For example, the first proposed draft for the IMO regulations for LNG as fuel was based on the experience of how to use IGC-Code requirements on the first LNG fuelled ferry MS Glutra (see Fig. 1).

The shortcoming of this approach is that the rules at the end fit perfect to similar cases but only to these cases. The overall idea is the aim to limit the room for interpretation, and, may be more importantly, the discussions to a minimum. Approval organizations like class societies and, may be more important, approval engineers are often in favour of this kind of regulations.

The counterpart to the prescriptive requirement is the **goal based requirement**. The suspicion of the proponents of prescriptive requirements always is that the supporters of goal based requirements have no clue or that they want to cheat. And vice versa. The author believe that to some extend these suspicions are justified.

A pure goal-based approach only formulates the goal of a requirement and may be on the level below the related functional requirements and that's it.[2] The rest is left to the designer and his client.

It must be noted that the formulation of safety goals and related functional safety requirements is the basic condition to reach a consistent safety level for different cases and technical solutions. This is also the only way to transfer safety requirements to new applications, to modify them in a responsible way and to do interpret prescriptive rules correctly.

Additionally goals and functional safety requirements are needed to make it possible to maintain the intention of prescriptive requirements even beyond the pension age of the authors of these requirements! Again the sizing of PRVs is a prominent example how difficult it is to understand the background of prescriptive requirements when the knowledge about this is not present any more (comp. Annex, p. 137).

Coincidentally, the author thinks that the structure of the IGF-Code is a good example of a compromise between goal based approach and prescriptive regulations. It might also be possible that this interpretation is related to the fact that the initial proposal was proposed to IMO by Germany and that the author worked as a consultant for the subject for the German ministry of transport at that time.

---

[2] Comp. also Sect. 4.1, p. 24.

1.2   The Structure of the IGF-Code

The IMO IGF-Code: "International Code of Safety for Ships using Gases or other Low-Flashpoint Fuels" [3] is the mandatory instrument for low *flashpoint* ship fuels.[3] The IGF-Code is part of SOLAS [4] and therefore international law.

The IGF-Code is written as a Goal based Standard according to the IMO understanding. The Code has two parts:

1. The mandatory Part A for **all** alternative fuels (low flashpoint fuels) includes the Chapters "2. General", "3. Goal and functional requirements", "4. General requirements".[4]
2. The mandatory Part A-1 for ships using natural gas as fuel.

The intention in developing Part A was to have a mandatory part for all alternative fuels. Fuels not covered by the current IGF-Code can be approved by use of the alternative design approach as given in Part A, Chapter "2.3 Alternative Design".

Part A should also be regarded by everyone to be valid for the fuels which are covered by the *interim guidelines* developed (Methanol/Ethanol, Fuel Cell Systems). They should also be regarded by everyone to be mandatory for the interim guidelines currently under development (LPG, low flashpoint oil fuels, hydrogen, ammonia). The interim guidelines are NOT mandatory. They represent draft versions of special chapters for the named fuels which might become mandatory like the Part A-1 of the current IGF-Code!

Note that the author use the term "...should be regarded by everyone..." instead of "is" on purpose. Currently the interim guidelines do not fully reflect the text of IGF-Code Part A completely. The author feel confident that latest at the point in time when the interim guideline become mandatory there will be a political discussion about whether Part A of the IGF-Code should be applied.

Part A-1 is split into two parts for each Chapter. The first part consist always of the Sections "*.1 Goal", "*.2 Functional requirements", "*.3 Regulations-General" which give the reference to Part A of the Code and provide guidance about the individual Chapters. This includes special functional requirements for the individual Chapters if needed. The following sections give the dedicated prescriptive rules.

By the way, this is the reason why all alternative fuels can be approved by using the first three parts of IGF-Code A-1 as a basis for an alternative design in addition to IGF-Code Part A. A guideline is not necessary in this case! Following an IMO Guideline allows to avoid a separate alternative design approval if the Administration permits.

---

[3] Nowadays (since IMO CCC-8) alternative ship fuels.

[4] The "Preamble" of the IGF-Code is not named Chap. 1. but it is because otherwise part A of the Code could not start with Chap. 2.

1.3   Conventional Oil Based Ship Fuels

Until the early 2000s, fuel for general shipping was Heavy Fuel Oil (HFO) with a high Sulphur content of more than 2 %(*mass*) and Marine Gas Oil (destilate fuel) (MGO), Marine Diesel Oil (mixture of mainly destilate fuel with heavy oil) (MDO) with a flashpoint of 50 °C and higher.

There are three reasons to use these fuels. For a long time, a two digit amount of the total crude oil could not be refined to higher hydrocarbons and was refinery waste. As a result, HFO was and still is relatively cheap.[5] To substitute coal by HFO for the steam boilers used for steam engines was therefore a good idea to safe money and to reduce the volume needed for ship fuel. In shipping piston engines became competitive with steam engines in the 1950*s*. The disadvantage of these engines was that they needed to run on distillate fuel (MGO, MDO) because the cheap HFO used for steam boilers was simply too "dirty" and damaged the engines.

With increasing piston engine efficiency in the 1950*s* the high price for MGO could be offset.[6] The question for the refinery industry was therefore: What to do with the refinery waste HFO? The commercial solution was simple. Piston engines needed to become fit for HFO. The technical pathway towards this solution was less simple. Anyhow, in the 1950*s* the refinery industry financed research work on the 2-stroke engine to enable the use of HFO in piston engines (comp. [5], p. 276, [6]). It is well known that the research was successful. In the 1960*s* also medium speed engines also used HFO. In the 1970*s* all main engines in the seagoing merchant fleet used HFO. The attack of the cleaner distillate fuels against HFO was defeated and high Sulphur HFO with more than 2 %(*mass*) Sulphur remained the dominating ship fuel until the Sulphur content was limited to 0, 50%(*mass*) in 2020.[7]

The conventional ship fuels HFO, MGO with their high flammability limit were regarded as well established with a good safety record. It has to be noted that engine room fires often caused by high temperature surfaces like exhaust piping or other engine components are the most frequent reasons for fire on board of ships. This was one reason that liquefied gases and liquids with flashpoints below 60 °C were regarded as less safe to handle compared to oil-based fuels with high flashpoints (comp. Glossary, p. 155).

The main measure to define the probability of ignition for ship fuels is the flashpoint of the fuel.[8] The general limitation is given by the requirement that "..., no...fuel with a

---

[5] Note that fuel for international shipping is not taxed and is difficult to tax because the ships operate worldwide.

[6] The technical reasons for higher piston engine efficiency compared to steam engines is an interesting subject but is not the subject of this book.

[7] Higher Sulphur content is only allowed if Sulphure Dioxide (SO_2) is removed from exhaust gases by scrubbers.

[8] SOLAS: Ch II-2: Construction–fire protection, fire detection and fire extinction, PART A–General, Regulation 3, Definitions, "24. Flashpoint is the temperature in degrees Celsius (closed cup test) at

flashpoint of less than 60 degrees C shall be used;..." [4].[9] The IGF-Code [1] opened the way for Low Flashpoint Fuel (LFF) for shipping. In addition the revision of the IGC-Code [2] in 2008/2010 opened the possibility to use other fuels than oil fuel also on gas carriers which are not LNG carriers.

Up until 2021 all proposed fuel alternatives had low flashpoints. Therefore, these fuels were referred to as low flashpoint fuels. With the proposal of Ammonia as ship fuel this definition was not valid for all alternatives to oil based fuels any more.[10] For this reason, the naming was changed from low flashpoint fuels to alternative fuels by the *IMO CCC*-8 in 2022.

1.4 Background on the Introduction of Methane as General Ship Fuel

The burning of high Sulphur fuels creates a high content of SO_2 in the exhaust gas and SO_2 may return to the soil as acid rain. The problem was mainly related to land based power generation by coal power plants. Shipping only contributed to the problem in coastal areas.

The Scandinavian countries, especially, had bad experiences with *air pollution* caused by emissions of SO_2, NO_x, Particulate Matter (PM) to air. SO_2 and NO_x emissions caused the acid rain which heavily affected the lakes in Scandinavian countries which have a low carbon content in the water and therefore very limited possibility to neutralize the acid rain.

In the 1970s the southerly neighbours of Denmark, Sweden, Norway, Finland in the south developed a creative way to handle air pollution in their countries. They built high chimneys and avoided acid rain according to their country's limits. Because of the direction of the prevailing wind on (from West and South/West) the Scandinavian countries ended up with the problem.

The last coal fired power plant in Germany which attempted to take advantage of this "emission reduction method" was the brown coal power plant Buschhaus in northern Germany which went into service in March 1985.[11] Buschhaus had a chimney of 307 m, the highest in Germany. It went out of service in September 2016. As far as the author is aware, nearly all coal fired power plants in Europe have desulphurization plants to clean the exhaust gases. Acid rain is no longer a problem in Europe.

The lesson for Scandinavian countries was that air pollution (comp. Glossary p. 155) must be limited. Norway engaged early in air pollution prevention even if the absolute effect of the engagement may be regarded as small. In 2007 a NO_x tax was introduced with 15 NOK per kilo NO_x [7]. Parallel to the introduction of the tax, a NO_x fund was established as

---

which a product will give off enough flammable vapour to be ignited, as determined by an approved flashpoint apparatus. " SOLAS: [4].

[9] SOLAS Ch II-2, Regulation 4 Probability of ignition, item 2.2.1.

[10] Flashpoint NH_3 is 132 °C.

[11] Desulphurization was implemented in 1987.

Fig. 1.1 Engine room on board of MS GLUTRA. (*Source* Dr. Gerhard Filip (2003), Exergetics GmbH, Eckenerstraße 50, D-88046 Friedrichshafen)

an incentive with the intention to use the tax money for NO_x reduction measures. A relevant contributor to the NO_x emissions was traffic. The public fjord ferry business contributed a notable share of all traffic pollutions. LNG was introduced as a ship fuel in order to reduce these emissions. With at least more than 80 % of CH_4, LNG is the cleanest fuel with respect to air pollution with NO_x and particles.

In 2000 the first LNG fuelled ferry MS GLUTRA[12] was built (see. Fig. 1).

She still is running in regular ferry service. The GLUTRA is equipped with four spark ignition gas engines from Mitsubishi with 675 kW nominal power installed in separate housings on the main deck [8] (comp. Fig. 1.1). The engine housing follows the Emergency Shut Down engine room concept according IGF-Code (2016), Sec. 5.6 (ESD-ER-Conc) (comp. Sect. 5.6, [1]).

The LNG is stored in two 32 m^3 vacuum perlite insulated tanks on both ends of the ship below the car deck [8].

The safety concept for the vessel based on experiences gained from LNG Carriers which are built according to the IGC-Code [2].

In 2004 Norway proposed to IMO to accept a Code for gas as ship fuel developed by Norway. This Code was based on the rules developed for GLUTRA and very prescriptive.

Very soon it became obvious that this jump was insufficient for a general Code for gas as ship fuel for ships other than gas carriers. Between 2004 and 2009 interim guidelines were developed [9] and the IGF-Code followed in 2015 (comp. [10]).

The original proposal from 2004, the 2009 interim guidelines and the 2015 IGF-Code were intentionally based on the safety principles developed for the bulk transport of liquefied gases [1].

---

[12] IMO No-9208461.

1.5 Alternative Ship Fuels Other than Methane

In the early $2000s$ air pollution was the motivation for IMO to work on alternatives to the common oil based fuels. Nowadays the reduction of *Green House Gas (GHG) emissions* is the main motivation.

The IGF-Code is still the covering international legislation for IMO regulations on alternative fuels.[13] By reference in SOLAS [11] the IGF-Code became international law. For this reason, LNG and Compressed Natural Gas (CNG) are the only alternative fuels with mandatory legal basis. For the relevance of this compare also p. vii.

The additional fuels regulated so far by an IMO instrument are CH_3OH, C_2H_5OH [12] in 2020 and Liquefied Petroleum Gas (mixture of Propane unad Butane) (LPG). These fuels are regulated by two interim guidelines. An interim guideline is no mandatory instrument. It is voluntary. Flag States can accept the guidelines as a basis for a SOLAS certificate but they do not have to do so. The weaker legal status of a guideline might also have implications for the ship insurance.

Table 1.1 lists the current status[14] of rule development for alternative fuels at IMO.

For a long time guidelines for low flashpoint Diesel fuel were on the IMO Agenda but without relevant progress. These interim guidelines should include oil fuels with a flashpoint between $52\,°C$ and $60\,°C$, and cover oil-based fossil fuels, synthetic fuels, biofuels and any mixture thereof. Considering the relevance low flashpoint Diesel hopefully may have as *Power to X Fischer Tropsch fuel (PtX-FT)* in future these guidelines are of great relevance.

Table 1.1 List of IMO rule development for alternative fuels in shipping

Natural gas	LNG	IGF-Code	In force since 2017	Mandatory international law
Propan/Butan Mixtures	LPG	Interim guideline	Adopted	Voluntary use if administration permits
Methanol and Ethanol	– –	Interim guideline	Adopted	Voluntary use if administration permits
Low falshpoint oil fuel, FT-Diesel	– –	Interim guideline	Under development	Could be used for FT-Diesel
Hydrogen	–	Interim guideline	Under development	–
Ammonia	–	Interim guideline	Under development	–

---

[13] comp. Sect. 1.2, p. 3.
[14] First quarter 2024.

Related to the discussion on the reduction of the CO_2 footprint in shipping recently Hydrogen and Ammonia got in focus as ship fuel. Currently interim Guidelines for both fuels are under development and there is a political pressure to develop them fast.

As explained in Sect. 1.2 Part A of the IGF-Code [1] can be used for all alternative fuels in shipping. In author considers that this is needed to ensure a consistent level of safety for all these fuels. The tendency to develop stand-alone regulations for different fuels without a traceable relation to the IGF-Code may result in different safety levels. This should be avoided.

The safety principles outlined in this publication are valid for all alternative fuels.

References

1. IMO (2016), IMO Resolution MSC.391(95), IGF-Code: International Code of Safety for Ships using Gases or other Low-Flashpoint Fuels, IMO, London, ISBN 978-92-801-1653-3
2. IMO (2016), IMO Resolution MSC.5(48), MSC.370(93), IMO IGC-Code, International Code for the Construction and Equipment of Ships Carrying Liquefied Gases in Bulk, IMO London, ISBN 978-92-801-1631-1
3. IMO (2015); RESOLUTION MSC.391(95) (adopted on 11 June 2015) ADOPTION OF THE INTERNATIONAL CODE OF SAFETY FOR SHIPS USING GASES OR OTHER LOW-FLASHPOINT FUELS (IGF CODE), London
4. SOLAS 2020 Consolidated Edition; ISBN: 978-92-801-1690-8, IMO, London
5. Wiborg, Susanne, Wiborg, Klaus, Dr. (1997); Unser Feld ist die Welt-150 Jahre Hapag-Lloyd, publisher Hapag-Lloyd AG, Hamburg
6. Rulfs, Horst, Prof. Dr. (2007); Marine Heavy Fuel Oils-Problems and Alternatives; Ship Efficiency Conference, Hamburg
7. Story about the NOx Fund https://www.noxfondet.no/en/articles/about-the-nox-fond/, Cited 28.04.2023
8. Einang, Per Magne; Haavik, Konrad Magnus; The Norwegian LNG Ferry; PAPER A-095 NGV 2000 YOKOHAMA
9. IMO (2009), IMO Resolution MSC.285(86) Interim Guidelines on Saftey for Natural Gas-Fuelled Engines, IMO, London
10. Wuersig, Gerd (2015); The development of THE INTERNATIONAL CODE OF SAFETY FOR SHIPS USING GASES OR OTHER LOW-FLASHPOINT FUELS (IGF CODE)-IMO MSC.391(95); Motorship Conference-Gas fuelled ships-Hamburg, 10/12 Nov. 2015
11. IMO (1974); International Convention for the Safety of Life at Seas (SOLAS), amended by resolution MSC392(95), IMO, London
12. IMO (2020); IMO Resolution MSC.1/Circ.1621; INTERIM GUIDELINES FOR THE SAFETY OF SHIPS USING METHYL/ETHYL ALCOHOL AS FUEL; IMO, London, 7 December 2020

Risk Evaluation

The IGF-Code requires that the *risk* shall be equivalent to the risk of conventional oil fuelled ships ([1], Sect. 3.2.1). The problem here is that an accepted quantification of the risk for conventional ships does not exist.

Risk is defined as a relation between the likelihood of an event occurs and the consequence this event will have if it occurs. The aim of Quantitative Risk Assessment (QRA) methods is to calculate a risk figure to enable exact quantitative risk judgement. The other approach to evaluate technical risk is a qualitative judgement. For both methods an idea about the frequency of occurrence and the severity of the consequences is needed and as shown below is possible and useful.

2.1   What is a Risk?

The most commonly used definition of technical risk (Ri) is that it the product of frequency (fr) and the consequence of occurrence (co) (Eq. 2.1)

$$Ri = fr \cdot co \tag{2.1}$$

© The Author(s), under exclusive license to Springer Nature Switzerland AG 2025
G. Würsig, *The Safety Principles for the Use of Low Flashpoint Fuels in Shipping*,
Synthesis Lectures on Ocean Systems Engineering,
https://doi.org/10.1007/978-3-031-64174-9_2

2.1.1 Frequency of Occurrence

The risk (Ri) is related to the frequency of occurrence (fr) and the consequences of the failure event (co). The frequency is defined either related to a time.[1] or to in relation to the number of operations.[2]

As shown in the following both definition can be converted into the other. The "normal" failure frequency[3] of a moderate complex technical system like e.g., a cheap TV set can be assumed to be approx. $1 \cdot 10^{-4}$ on request. This means that the TV set fails in 1 out of 10.000,–requests to operate.[4]

Assuming that the TV-set is switched on 2 times a day during one year the number of requests become 730 per year. The average time between failure therefore is $10.000,-/730 = 13,70$ years. The yearly failure rate become $1/13,70 = 0,073$ or $7,3 \cdot 10^{-2}$ per year.

For the risk evaluation it is not relevant to know which component prevents the TV set from working. The failure rate of 1 in 13, 70 years is also no guarantee that the TV-set runs without problem for nearly 14,–years. It might fail long before or after. At least the TV-set manufacturer trust his product for 1460 on/off cycles or 14, 6% of the failure rate within a 2 year guarantee period.

It could be argued, that people operate their TV with a much higher frequency. Doubling the frequency to 4 times a day will result in only approx 7 years between failures. On the other hand increasing the reliability by reducing the failure rate to $0,5 \cdot 10^{-4} = 5 \cdot 10^{-5}$ will give the same result as before.

The simple example demonstrate that the failure frequency is a good tool to indicate the likelihood of an event. At the same time it demonstrates that the uncertainties included in the failure rate make it useful for risk evaluations to work with ranges instead of fixed values.

To define a practical range for failure categories in safety analysis where no detailed failure statistics are available is to consider a range of 1–5 as one range and 6 to 10 as the next higher range (see Table 2.1, p. 11):

For the example of the TV set this will be:

For the individual TV set 1 failure per year (range: Frequent) is clearly to much and 1 failure every 1000,–(range: Rare) years would be nice but is not likely.[5]

Whether a failure frequency is tolerable or not, also depends on the number of units in operation. Therefore, the final judgement related to qualitative failure considerations

---

[1] E.g., occurrences in one year of operation.

[2] E.g., occurrence of a failed start of an engine related to the number of start/stop operations (failure frequency on request).

[3] Failure rate.

[4] Note that this definition is related to an switch on shut off cycle because to shwitch on the TV one must have shut off it before.

[5] To use 1.369, 9 instead of 1.000,–would mean an accuracy which is not covered by the simplified failure model used.

Table 2.1 Practical definition of failure ranges for qualitative risk assessments (failure on request)

	Failure on request	Failure on request
Frequent	$1\text{--}5 \cdot 10^{-3}$ and $6\text{--}9 \cdot 10^{-4}$	Range: 10^{-3}
Moderate	$1\text{--}5 \cdot 10^{-4}$ and $6\text{--}9 \cdot 10^{-5}$	Range 10^{-4}
Seldom	$1\text{--}5 \cdot 10^{-5}$ and $6\text{--}9 \cdot 10^{-6}$	Range: 10^{-5}
Rare	$1\text{--}5 \cdot 10^{-6}$ and $6\text{--}9 \cdot 10^{-7}$	Range: 10^{-6}

includes the number of units in operation. For the TV set the "Moderate" failure rate of 10^{-4} on request for a single unit (Table 2.1) corresponds to a range of 10^{-1}/year. This time related figure can be multiplied by the number of units to judge the acceptability of the failure rate.

$$uy = fr \cdot n$$

The interpretation is that the failure rate of $fr = 10^{-1}$/year corresponds to 1 out of $n = 10$ units which will fail within one year ($uy = 1$). If this is sufficient depend on the risk related to the failure. For the TV set the risk for the manufacturer is that he has to repair the set on guarantee. Here the above thoughts may become useful for him.

2.1.2 From Failure Frequency to Risk

The judgement of the acceptability of the risk requires the consideration of the consequences. For the TV set example, the consequences can be related to money which is the easiest way to define a consequence. The consequence for the individual user is that he must buy a new TV set which might cost e.g. 400, − $. Using Eq. 2.1 the risk for the TV user become

$$Ri = 0,073 \cdot 400, - = 29,2 \text{ \$/year}; \quad fr = 0,073 \text{ /year}, \quad co = 400, -\$$$

Applying the risk range ($fr = 10^{-1}$) for the "Moderate" failure rate in Table 2.2 will give a risk of $Ri = 0,1 \cdot 400 = 40$ \$/year instead of 29 \$/year.

The view of the manufacturer has not to consider the price of the TV set as consequence but the repair costs within the guarantee period related to the price of the TV set. The number of units is therefore also of interest to the manufacturer. From the above the manufacturer has to consider that during 1 year guarantee period 0,073 TV sets need a repair under guarantee. Assuming costs of $co = 50$ \$ for the repair the risk becomes

$$Ri = 0,073 \cdot 50 = 3,65 \text{ \$/year}; \quad fr = 0,073 \text{ /year}, \quad co = 50\$$$

Table 2.2 Illustration of the failure ranges for the TV example

	Index	Failure rate	Years between failure	Range fr: 1/year
Frequent	1	$730 \cdot 10^{-3} = 0,73$	$1/0,73 = 1,37$	1
Moderate	2	$730 \cdot 10^{-4} = 0,073$	$1/0,073 = 13,7$	10^{-1}
Seldom	3	$730 \cdot 10^{-5} = 0,0073$	$1/0,0073 = 137,0$	10^{-2}
Rare	4	$730 \cdot 10^{-6} = 0,00073$	$1/0,00073 = 1.369,9$	10^{-3}
Not considered	5	$730 \cdot 10^{-7} = 0,000073$	$1/0,000073 = 13.698,6$	10^{-4}

Working with the risk category instead of the more exact figure results in $Ri = 0,1 \cdot 50 = 5$ \$/year. The risk for the manufacturer therefore is $3,65\$/400\$$ or $0,9\%$ of the sales price.[6]

In the simple example above the failure rate is high because it must be expected that $7,3$ ot of 100 units will need repair under guarantee within a one-year guarantee period. On the other hand, the related costs are not so high. $0,9\%$ of the sales price may be acceptable.

The simple TV set example illustrates how useful risk evaluations are for decision making and how different a risk can be for the same case but different groups which are subject to the consequence of a failure.

In technical risk evaluation the need to define consequences is often not only related to money but to human health and also to environmental damage. The quantification by giving these consequences a price tack is to enable a ranking is often done but problematical. What is the value of health and death? For the individual the own health has a very high value and the death is a no go! For a life insurance company the likelihood and value of death is the base for the insurance contribution.

For a useful application of QRA and also Semi-Quantitative Risk Assessment (S-QRA) a Qualitative Risk Ranking (QRR) is necessary. A useful approach to enable a ranking is to describe the consequences and to relate a numeric figure to them. This figure might be failure frequencies and monetary cost of the consequence of failure. Very often index figures are used. Table 2.3 gives a possible definition for consequences using an index figure for a S-QRA. This approach is widely used in industry in different versions and with different definitions.

Note that fatalities are ranked very high in this consequence ranking. It is also usual to accept a limited number of fatalities in the major effect category and to stipulate a high fatality rate in the hazardous effect category.[7] If such a ranking is related to ships, "system" should be substituted with "ship" in Table 2.3.

---

[6] $1,25\%$ when using the risk category.

[7] Whatever is is regarded as "limited" and "high" fatality rate!

Table 2.3 Definition of consequences for a technical risk assessment

Consequence	SE-Index	Effect
No effect to normal system operation	1	No
Disturbed system operation	2	Minor
System shut off, repair needed	3	Moderate
Human injuries and/or severe damage or loss of system	4	Major
Fatalities, loss of system, and/or damage to/loss of other systems	5	Hazardous

2.2 Risk Ranking

The above sections describe the basics of a possible method to define the frequency of failure occurrence and the definition of the consequences for technical systems like the systems relevant here for the application of alternative fuels in shipping or for industry process systems in general. The missing link is the risk ranking as a combination of frequency and consequence as defined by Eq. 2.1.

The widely applied solution is the definition of a risk matrix which compiles the data of *HAZard IDentification study (HAZID)* and *Failure Mode and Effect Analysis (FMEA)* results.

Using the index for the failure frequency according to Table 2.2 and combining it with consequence index from Table 2.3 gives a risk index which is often calculated by simple multiplication (Eq. 2.2).

$$Ri = fr \cdot co$$

Table 2.4 gives the ranking matrix.

Most readers may know this matrix in the green/yellow/red coloured version and most times without the Ri number. The colouring is related to the risk potential of a system,[8] the related detailed definitions of failure effects and the judgement on failure consequences which need to be defined carefully. Therefore the principle matrix given by Table 2.4 is not coloured.

In the early 2000s the author introduced the risk ranking methodology described above based on HAZID, FMEA methods in the company he worked for. To the knowledge of the author the method is still applied there.

---

[8] Ranging from TV sets to nuclear power plant.

Table 2.4 Principle risk matrix

Effect	Index consequence	Not possible	Rare	Seldom	Moderate	Frequent
		1	2	3	4	5
		Risk index (Ri)				
No	1	1	2	3	4	5
Minor	2	2	4	6	8	10
Moderate	3	3	6	9	12	15
Major	4	4	8	12	16	20
Hazardous	5	5	10	15	20	25

2.3 The Relevance of the Number of Units in Operation

From the illustration of failures for a TV set the relevance of units in operation have been illustrated (comp. p. 11). The unit effect becomes most relevant for systems which might have relevant effects on third parties in case of failure.

The current and past risk for loss of life in commercial passenger flight business may illustrate the unit effect because airline statistics illustrate the success story of a drastic risk mitigation in passenger flight business.

According to the International Air Transport Association (IATA) figures the passenger related risk for loss of life was $6,23364 \cdot 10^{-8}$/year in 2019. This corresponds to approx. 280 loss of life events out of $4,5 \cdot 10^9$ passengers.

In 1970 the risk was at $378,788 \cdot 10^{-8}$/year related to 1.218 losses of life at a transport capacity of only $0,32 \cdot 10^9$ passengers per year. With the passenger number from 2019 and the risk figure from 1970 there would have been more than 17.000,–fatalities in 2019 or the fatal crash of nearly 3 passenger jets per week[9]! This example illustrates the considerable difference the number of units make when talking about risks. It also illustrates the limits of risk mitigation.

The message from the above for the use of alternative fuels in shipping which might have an increased risk potential is simply that it makes a big difference if some hundred units are in operation within a limited, well controlled industry like the LNG transport or if thousands of units operate in general shipping with LNG as fuel. The author was, is and most likely will be persuaded that the benefits of operating ships with alternative fuels like LNG are much higher than the risk created by most of these fuels. Nevertheless, a risk will always remain at a minimum level which at best might be close to what we see currently in passenger flight business.

---

[9] Assuming 120 passengers per plane.

This book is not a book about quantitative risk mitigation. It is a book about the technical safety principles to apply alternative fuels in shipping in a responsible manner. Therefore, the following remarks on the thoughts about failure frequency, failure consequences and the risk as a result of both may complete the subject for this book and should remain as a background for further reading.

As a rule of thumb the technical failure rate of a moderately complex technical system can be assumed to be approx. 10^{-4} on request (for an illustration comp. above). The author considers that a maximum achievable safety level against major, hazardous events from using alternative fuels in shipping should be approx. 10^{-6} per year for major events and approx. 10^{-7} per year for hazardous events. The technical mitigation and also, to a lower extend, the administrative measures must ensure the bridge between the failure rate of a single component and the failure rate with major or hazardous consequences.

For a number of approx. 70.000,–seagoing ships the above figures would therefore limit the number of major events related to alternative fuels to about 1 within 15 years and to hazardous events to 1 within 150 years if all ships run on alternative fuels.

Reference

1. IMO (2016), IMO Resolution MSC.391(95), IGF-Code: International Code of Safety for Ships using Gases or other Low-Flashpoint Fuels, IMO, London, ISBN 978-92-801-1653-3

Design Principles for Alternative Fuels Systems for Seagoing Ships

3

Technical requirements for process and propulsion systems on seagoing ships are similar to those for industrial land based installations because the safety considerations are the same. Nevertheless, often engineers who are used to design land based process technology applications wonder about "strange" requirements for ships which appear to them unnecessary. For a better understanding, the special situations on board of ships should be considered.

3.1   Why Ships are Different

No one designs process or propulsion technology especially for ships. The simple reason is that land based applications for equipment used on board is, in nearly all cases, the much larger market. Exemptions from this rule are very limited. This is not very often recognized and considered by the shipping community.

Ships are not mass products. This is the case even if it is considered that ship even of the same types have limitations in the design variations and that series of ships[1] are common. On the other hand, the process- and propulsion system components are mass products.

Some items which make ships different are:

- Ships are watertight structures with limited possibilities to the unforced ventilation of spaces. The risk of accumulation of dangerous substances is therefore higher than for land based process systems.
- Ships move and have a very high potential energy due to their speed and weight. The likelihood of collisions can be reduced but cannot be excluded.

---

[1] E.g. approx. 10 Container ships for an Asia/Europe trade.

© The Author(s), under exclusive license to Springer Nature Switzerland AG 2025
G. Würsig, *The Safety Principles for the Use of Low Flashpoint Fuels in Shipping*,
Synthesis Lectures on Ocean Systems Engineering,
https://doi.org/10.1007/978-3-031-64174-9_3

- Ships are exposed to vibrations from technical equipment and forces related to the movements caused by the sea state. For this reason, e.g. fully welded pipes can not be regarded to be technical tight under all conditions and at any time as it is assumed for land based process systems.
- Ship equipment must be failure tolerant under any operation circumstances. The sufficient energy supply in all situations is of high relevance. E.g., it is not acceptable to lose the propulsion power in a storm.
- In general, ships are designed for 25 years of operation. The real life of a ship may be much longer. E.g., ferries and cruise ship often run much more than 25 years.
- Ships are operated by a large number of legal entities on an international basis. These entities have different safety cultures, maintenance strategies and different views on technical responsibility.
- The escape possibilities from a ship are limited and risky. E.g., if a car catches fire you can stop the car and run away. On a ship at open sea, may be in a storm this gets difficult.
- The definition, application and maintenance of technical safety standards on a worldwide basis is more consistent for ships compared to shore site applications.

  - The IMO regulations ensure a consistent definition of the necessary basic safety standard for all ships.
  - The Flag State administrations are responsible to ensure that ships are built according the applicable standards and that these standards are maintained over the lifetime of the ship.
  - The class societies survey the rule compliance of design,[2] building[3] and operation.[4] At the same time they work to a large extent on behalf of the Flag States to ensure that the Flag State requirements are fulfilled. The scope of work and responsibility of Class Societies is much larger compared to shore approval organisations like the Technischer Überwachungsverein (TUV) or the Deutscher Kraftfahrzeug-Überwachungs-Verein (DEKRA).

- Leaks during fuel transfer[5] are the most likely reason for the release of large amounts of fuel.

  - Fuel bunkering involves much larger transfer amounts of fuel than most land based fuel supply operations. E.g. LNG fuel supply for a small Container feeder vessel may be between some $300, - m^3$ and $2.000, - m^3$. The maximum amount is related to

---

[2] Class approval of design drawings.

[3] Survey of building process at the yard.

[4] Continuous technical survey of safety relevant ship technical items (5-years class run).

[5] Bunkering.

large Container Ships which bunker approx. 18.000, – m^3 LNG for one round trip of about 100,–days.[6]
- Fuel transfer is subject to movements arising from the sea state[7] and possible tidal changes.[8]

3.2   Separation of Process and Fuel Storage Systems from the General Ship Arrangement

For crew and passengers, the ship is a working and living space. On Gas Carriers there is a strong separation between the storage and handling space for liquefied gases and the living space of the crew by defining a *cargo area* which is separated from the rest of the ship. In general, the cargo area includes all spaces with cargo tanks and cargo related process equipment. This is the main part of the ship excluding the deck house and the bow area where the anchors and bow mooring winches are installed. The deck house and aft of the ship are typically located behind the cargo area.[9] In all cases they are clearly separated from the cargo area. For the cargo area strict explosion protection and EX-Zoning applies.

The same separation principle is applied for shore site chemical and refinery systems and on oil-, gas industry offshore applications.

It should be noted that the strong physical separation on board of gas carriers and also offshore installations has limitations. E.g. on LNG gas carriers the need to supply cargo BOG to the engines implies that gas must be transferred below the deck house into the engine room. The two barrier principle is used for separation of gas dangerous space and safe spaces in these areas (comp. Chap. 5, p. 29). The level of safety is regarded to be equivalent to the physical separation between cargo area and deck house as described above. In most cases, the equivalent level of safety is not quantified if the separation regulations are fulfilled.

For general cargo and passenger ships using Low Flashpoint Fuel (LFF), the two barrier principle (comp. Chap. 5) is the main safety tool to reach the "saftey equivalence" stipulated in the IGF-Code, Sect. 3.2.1. How this principle is applied correctly is often difficult and a major challenge for the designer and approval authorities.

The challenges is increased by the commercial considerations which are a basis for any project for yard, ship owner, financing. There is always competition between different design proposal. There is a the risk that designers who do not know about the reason behind proposed two barrier designs present commercially very attractive designs and win the contract. The

---

[6] Note that the fuel amount with the same energy contend will be 2, 5 times these values if Liquefied Hydrogen is used as fuel.

[7] ship to ship bunkering

[8] Shore to ship bunkering.

[9] Some gas carriers have the deck house before the cargo area at the bow.

basic principle that a Type-C tank is not only a pressure vessel is a prominent example for such lack of knowledge (comp. Sect. 6.5, p. 46)!

3.3   There are No Tight Systems in the World

During the author's mechanical engineering studies at Leibniz University in Hannover in the 1980 he attend a lecture on heat exchanger design given by Prof. Buxmann.[10] Prof. Buxmann was an authority which was emphasized by his physical presence.[11] During the explanation of a large condenser installed downstream a turbine in a nuclear power plant a student asked what the component illustrated below the condenser might be?

Prof. Buxmann explained that this is the drip tray for leaks. The student was not happy with this answer and asked why a drip tray is necessary? A leak should be impossible because the heat exchanger is part of a nuclear power plant. Prof. Buxmann turned around to the student, looked at him and said "There are no tight systems in the world and this is the drip tray". He turned around to the picture and continued the lecture. The discussion was finished. Since that day the author knows and experienced later that there are "no tight systems in the world".

This statement could conclude this Section but first of all there are a lot of people who do not believe this and second there is a need to have "tight systems" or at least "sufficiently tight systems" like e.g. the condenser mentioned above. For ship safety, sufficiently tight systems are needed to limit the risk of asphyxiation, toxic-, ignitable atmosphere.

A current example of violating the above principle is the proposal to regard fully welded pipes for liquefied gas as ship fuel as "tight" and therefore not requiring a secondary barrier or a special control like gas detection when routed to enclosed spaces. This opinion comes from industry applications where gas installations run in large, well ventilated halls which normally do not move and in these cases it is a correct assumption. In ships it might become dangerous. However, it appears that this view is getting a majority one.[12]

The opposite "opinion" with regard to tightness is the fairytale that Hydrogen is passing in safety relevant amounts through all materials. None of the proponents naming this could ever explain to the author how a vacuum super-insulated hydrogen tank like the small one illustrated in Fig. 12.5, p. 104 could work when the hydrogen goes through the tank walls?[13]

Of course it is possible to design and build sufficiently tight hydrogen systems. Companies like Linde, Air Liquide and Air Products do this every day and of course systems designed for e.g. LPG will not be hydrogen tight!

---

[10] Prof. Dr.-Ing. Joachim Buxmann, born 19. August 1933, died 27. December 1996.

[11] At least this was the impression the author had.

[12] ...until the first systems will get untied with severe consequences.

[13] Please note that it is not possible to maintain the required vacuum of approx. 10^{-7} *bar* for the multilayer insulation by permanent evacuation.

The concept of ventilation during operation is closely related to the knowledge that there are no tight systems in the world. Unavoidable small releases are simply vented away to ensure reliable and safe system operation.

The opposite concept to ventilation is to have a surrounding with higher or lower pressure than the system pressure. A suitable solution, for example, is a double walled pipe with nitrogen overpressure in the space between the pipes. A more sensitive system will be a double walled piping with moderate vacuum. This system is definitely an option for heat sensitive cryogenic liquefied hydrogen where a nitrogen filling is simply not possible because the nitrogen would be solid at LH2 temperature.[14]

A gas tight enclosure, possibly evacuated, is a solution for piping but not for process equipment with valves, flanges etc. Such systems surrounded by a gas tight enclosure will most likely prove the above statement that "there are no tight systems in the world". Nevertheless, they are repeatedly promoted until they prove to be unreliable or extremely maintenance sensitive.

3.4   Boundaries to Limiting the Risk of Collision

The requirements and background to limit the risk of collision to an acceptable level according to the IGC-Code, IGF-Code [1, 2] are detailed in Chap. 8, p. 71 ff. In general it can be distinguished between the location[15] of collisions and the role of people effected by such an event.

At open sea and in limited waters (coastal waters, channels, rivers, harbour approaches) the speed of the ships are often high enough to cause damages of fuel or cargo tanks in the case of a collision. The most critical scenario is a T-Bow side collision. Aft collisions are seldom and bow collisions are less likely to damage the tanks. The distinction between gas carriers and gas fuelled ships is also relevant because a gas carrier is to at least 70–90% of its length a "cargo tank". Any side collision at open sea or in limited waters has a high probability to affect a cargo tank. This is not the case for gas fuelled ships because the extension of the tank related to the ship length is much smaller. In most cases it is approx 10% of the ship length or below.

Related to the relatively large distance form other objects the effect of an open sea collision is locally and the effect of parties not involved in such an event is limited. At least if a cloud of released liquefied gas and air can not reach populated areas until it is mixed below the Lower Explosion Limit (LEL)[16] or in case of NH_3 below the toxicity limit.[17] For this reason

---

[14] Note that Helium instead of Nitrogen is not a feasible solution for mass application of LH2 systems. Helium is simply too rare to waste it on such applications!

[15] With this, the related speeds and resulting energies released.

[16] $4\% = 40.000, -ppm$ volume for CH_4.

[17] Severe health effects: ALG-3, 10 min exposure limit $0, 27\% = 2.700, -ppm$ volume for NH_3 (comp. [3], Annex, Table 13 on p. 13).

it has to be distinguished between open sea collisions and in limited waters collisions. E.g., a sever collision involving a LNG gas carrier with a damage of an LNG tank on the river Elbe might have much higher consequences for people and installations on shore than it would have on the open sea.

The situation in port is less risky with regard to collisions because the possible speeds are limited. On the other hand, effects to the surroundings can be much more severe. Again distinction between non toxic, toxic cloud density below ambient air, above than ambient air is of relevance. In general it can be concluded that the structural protection of liquefied gas carriers and also gas fuelled ships during port operation is very high (comp. Chap. 8, p. 71 ff.) and therefore the likelihood of cargo or fuel release as a consequence of a collision in port is low.

It should be noted that any collision which damages the cargo or fuel tank will have liquefied gas release to the ambient as a consequence.[18] Fire and explosion will not occur in every case. As a conclusion from accident statistics and safety analysis the author assumes that fire and explosion may occur in 10–20% of collisions which damage the cargo/fuel tank. Please note that others stipulate that fire and explosion will occur in 100 % of collision events with liquefied gas release. These worst case scenario often is combined with the assumption that 100 % of the people in the surrounding will dy or will be severely insured. Experiences from ship incidents do not support this assumption. Also here a low two digit percent value is more realistic.

Finally, the question who will be affected by a collision should be considered. The basic distinction is between people professionally involved in the ship operation and, e.g., passengers on a ferry or a cruise ship who have nothing to do with the operation of the ship. Here the same distinction needs to be made, as it is done for risk evaluations of industrial sites, between people professionally involved in site operation and the general public. Professionals carry a higher risk because they are or at least should be aware of the risk and take it intentionally.

References

1. IMO (2016), IMO Resolution MSC.5(48), MSC.370(93), IMO IGC-Code, International Code for the Construction and Equipment of Ships Carrying Liquefied Gases in Bulk, IMO London, ISBN 978-92-801-1631-1
2. IMO (2016), IMO Resolution MSC.391(95), IGF-Code: International Code of Safety for Ships using Gases or other Low-Flashpoint Fuels, IMO, London, ISBN 978-92-801-1653-3
3. IMO submission CCC 9/3/2, Republic of Korea, 20 2023; AMENDMENTS TO THE IGF CODE AND DEVELOPMENT OF GUIDELINES FOR ALTERNATIVE FUELS AND RELATED TECHNOLOGIES, Study on Safety Assessment of Ammonia Toxicity, IMO, London 2023

---

[18] Of course only when using liquefied gas as cargo or fuel.

The General Applicable Principles in Part A of the IGF-Code

4

Part A of the IGF-Code [1], which include Sect. 2., 3., 4., is the general part and relates to all alternative fuels. Only Part A-1 (Sect. 5. ff) is limited to natural gas as fuel even if it can be applied to a large extent to other alternative fuels. The structure of the Code and the resulting general interpretation are described in Sect. 1.2, p. 3.

Part A can and should be applied to other alternative fuels and Part A-1 should be the base line for the level of prescriptive requirements for individual fuels. The current status of application for these fuels is given in Table 1.1, p. 7. The best way to ensure a consistent safety level of guidelines would be to make references to the IGF-Code [1] instead of repeating requirements without a clear reference track to the origin in the different guidelines.

This principle is currently not consequently practised kept in the development of the different guidelines. Most of the basic initial proposals are closely related to the current IGF-Code Part A and Part A-1 but the wording is already partly modified, and no reference indication is given. This is the reason that during the work on the guidelines the "wheel" is invented again and again. It must be considered that the development teams in Correspondence Groups and Working Groups are vary, the contributors have different backgrounds and partly also interests and understanding of what the right level of safety should be for "their fuel". Therefore, the different versions of the "wheels" look different and not all of them have a circular shape. This inconsistency lead to different safety levels for different fuels and might be one of the most critical items in rule development for alternative fuels.

In the following some most relevant items of IGF-Code part A are highlighted. The references are given to the original version of the Code [2] as adopted in June 2015 by IMO Marine Safety Committee (MSC).

© The Author(s), under exclusive license to Springer Nature Switzerland AG 2025   23
G. Würsig, *The Safety Principles for the Use of Low Flashpoint Fuels in Shipping*,
Synthesis Lectures on Ocean Systems Engineering,
https://doi.org/10.1007/978-3-031-64174-9_4

4.1   Goal and Functional Requirements of IGF-Code Part A

The baseline for safety requirements is given by the Goal and the Functional Requirements for all alternative fuels which should be fulfilled.[1] Deviations from this baseline
should clearly be explained and referenced. This should be done knowing that the Functional Requirements (FRs) are to some extent the intentions, wishes and hopes of the group
which developed them. Knowing this gives a reader the possibility to make an interpretation
of the background for sometimes not very self explanatory prescriptive requirements.

It should be noted that special FRs relevant for the different sections of Part A-1 for
LNG are referenced in the beginning of each section in Part A-1. It should also be noted
that the first three Sub-Sections of each Section in Part A-1 are intended to be applied for
other alternative fuels also. Therefore, they always define a Section Goal, special FRs and
General Requirements. For this reason these first Sections for each item are the baseline for
any alternative fuel (comp. also Sect. 1.2, p. 3).

The goal of the IGF-Code is "...to provide for safe and environmentally-friendly design,
construction and operation of ships...using gas or low-flashpoint fuel as fuel".[2] Even if the
main aim of the IGF-Code is ship safety the goal to be environmentally friendly led, e.g., to
the consequence that for LNG as fuel the release of any Methane must be avoided in normal
operation.[3] This aim is also specified in Functional Requirement (FR) Sect. 3.2.9 requiring
"...the system shall be designed to prevent venting under all normal operation conditions
including idle periods."

FR Sect. 3.2.1 [1] defines a high ambition even if the general mishap of this ambition is
that it is not quantified for any fuel used in shipping. It quotes: "The safety, reliability and
dependability of the systems shall be equivalent to that achieved with new and comparable
conventional oil-fuelled main and auxiliary machinery."

It is assumed that the FR is fulfilled if the prescriptive requirements in the fuel type related
part of the IGF-Code are fulfilled (for LNG: Part A1 of IGF-Code). The judgement on this
is, in general, not quantitative but qualitative.[4]

A second most relevant FR is given by Sect. 3.2.7 [1] requiring that "System components
shall be protected against external damages." The most relevant regulations related to this
FR are the location of system components in the ship to protect them from collision and
grounding damages. For the practical application of the FR, compare Chap. 8, p. 71. The
related regulations in Part A-1 for LNG[5] of the IGF-Code should be used as a baseline for
all alternative fuels.

---

[1] Section 3 of IGF-Code.

[2] Section 3.1 Goal [1].

[3] Section 6.9.1.1 IGF-Code "...maintaining tank pressure below the set pressure of...relief valves for a
period of 15 days.."; Sect. 6.9.1.2 "Venting of fuel vapour for control of tank pressure is not acceptable
except in emergency situations."

[4] For the practical application of the term "risk" compare Chap. 2, p. 9.

[5] IGF-Code Sect. 5.3 and specially 5.3.4

This part of the IGF-Code Part A-1 was the subject of a very controversial discussion for at least three WG meetings.[6] If used correctly the regulations in IGF-Code Sect. 5.3.4 are regarded by the author as a good compromise between the intention to apply an alternative fuel and the aim to limit the risk to an adequate level (comp. Chap. 8, p. 71). For this reason they should be the baseline for all alternative fuels.

For a ship the loss of power is one of the major concerns which must be avoided. An important reason is that heavy weather conditions and loss of power are frequently related to each other and in this situation the complete ship is at high risk. The principle that even fire and explosion events should not lead to an "...unacceptable loss of power..." is therefore stipulated in FR Sect. 3.2.12.

Finally FR Sect. 3.2.18 requires redundancy but also limits the redundancy to a single failure by stating: "A single failure in a technical system or component shall not lead to an unsafe or unreliable situation". The author regards this single failure principle as sufficient for most alternative fuels including Hydrogen. Considering the possible consequences of Ammonia release into the ship, the author does not think that it is adequate for all systems related to Ammonia as ship fuel.

4.2   Risk Assessment and Limitation of Explosion Consequences of IGF-Code Part A

Today large parts of industry, approval organisations and regulatory authorities believe in risk assessments and especially in so-called quantitative risk assessments (comp. also Chap. 2, p. 9). Some parties assume that with a risk assessment nearly no prescriptive regulation is necessary any more. On the other hand, it is accepted that technical designs approved in practise and repeated in a new application do not need a risk assessment again and again. For this reason Sect. 4.3 IGF-Code requires a risk assessment but it limits the scope to the following parts:

- Sizing of drip trays (Sect. 5.10.5 [1])
- Airlock design and application (Sect. 5.12.3 [1])
- Fuel containment system (Sect. 6.4.1.1 [1])
- Additional relevant accidental scenarios for type B independent tanks (Sect. 6.4.15.4.7.2 [1])
- Location of bunkering stations (Sect. 8.3.1.1 [1]) including those located inside of the ship (Sect. 13.7 [1])
- Alternatives for air exchange and related rates for tank connection spaces (Sect. 13.4.1 [1])
- Gas detectors at ventilation inlets to accommodations, machinery spaces (Sect. 15.8.1.10 [1])

---

[6] Working Groups are set up by IMO committees to work on a subject during IMO meetings (WG).

- Annex to Part A-1 IGF-Code on limit state design of novel containment system designs:

  – risk assessment required for possible structural tank failure (Sect. 4.4 [1]) and for additional accident scenarios (Sect. 6.8 of the Annex [1])

The dedicated definition of the requirements where an additional look with a risk assessment is useful to improve safety is regarded to be a good practical approach. Of course, the resulting list can and should be discussed for each alternative fuel individually. The items named above should be used as a guidance.

4.3   Is an Explosion the Same as a Deflagration?

To explain the background of Sect. 4.3 "Limitation of explosion consequences" in the IGF–Code it is essential to understand that the term "explosion" is defined as "... deflagration event of uncontrolled combustion" (Sect. 2.2.13 of [2]). A *deflagration* is characterised by a flame velocity below the sound velocity of the fuel and a pressure increase in the *mbar* range. In contrast to this a *detonation* is characterised by a high pressure increase to 8 *bar* or above and a flame velocity above the sonic velocity of the fuel. The term *explosions* often summarize deflagration and detonations.

The difference between "detonation" and "deflagration" was discussed intensively during the development of the Code because on the first glance there is no difference for most experts. At the end it became clear to the experts who developed the Code that there is a big and relevant difference. It also became clear that a detonation event cannot be controlled and simply must be avoided. The Code therefore only covers mitigations of the consequences of deflagration events! For this reason, the above definition is given. Let us wait and see how long it will take before someone has the idea to simplify the Code by deleting this definition.[7]

A deflagration in an open space is related to a limited pressure increase of some *mbar*.[8] At this pressure increase ordinary window glass is destroyed. Nearly all ignitions outside of an enclosure are deflagrations. Deflagrations in confined spaces can be mitigated by adequate pressure release. Detonations cannot be mitigated by pressure release.

Without a pressure release the maximum static pressure increase for Methane/Air or Hydrogen/Air mixtures can reach a factor of approx. 8 for stoichiometric mixtures. In other words such a deflagration in a confined space without pressure release occurring at atmospheric pressure of 1 *bar* may have a pressure increase to 8 *bar*.[9] Note that rupture of

---

[7] If no one will ever have the idea to delete it because this book is known this would be enough motivation for the author to write it.

[8] Less than 0, 07 *bar*[3], Sect. 6., p. 13.

[9] Gasoline/Air mixture deflagrations in confined spaces may reach approx. 5 to 6 *bar*.

human ear drum occur at approx. $0, 35$ *bar* and that not reinforced walls of houses may fail at approx. $0, 55$ *bar* [3] (p. 13, Sect. 6). Pressure release reduces the maximum pressure of a deflagration in confined spaces.

The careful reader may have noted that the author limited deflagrations in open space to "nearly all" events. In fact, there is only one exception known to the author. This is related to the jet release of hydrogen into semi-confined structures in open space like the pipework in a refinery, on a ship, or a tunnel. In such gases a strong ignition source might lead to a detonation event! This is one of the few differences between Hydrogen and Methane as an alternative fuel.

References

1. IMO (2016), IMO Resolution MSC.391(95), IGF-Code: International Code of Safety for Ships using Gases or other Low-Flashpoint Fuels, IMO, London, ISBN 978-92-801-1653-3
2. IMO (2015); RESOLUTION MSC.391(95) (adopted on 11 June 2015) ADOPTION OF THE INTERNATIONAL CODE OF SAFETY FOR SHIPS USING GASES OR OTHER LOW-FLASHPOINT FUELS (IGF CODE), London
3. J. Hord (1976), Is Hydrogen Safe, National Bureau of Standards, Department of Commerce, Washington, D.C. 20234, USA

The Two-Barrier Principle

5

IGF-Code [1]: "3.2.18 A single failure in a technical system or component shall not lead to an unsafe or unreliable situation."

This FR stipulates that multi failure scenarios are not considered and that the protection against a single failure is sufficient. This is the motivation for the two-barrier principle.[1] The 2-barrier principle is a general concept to ensure that a single failure will not lead to an unacceptable dangerous situation. This generic concept is one of the most important principles for an adequate level of safety for system design.

The function of the secondary barrier must be controlled to ensure that it is available if needed. E.g. a double wall piping without monitoring of the integrity of the outer pipe is not following the 2-barrier principle because it cannot be ensured that the second barrier is working if the first barrier fails.

5.1   Double Walled Piping and Pipe Channels

Double walled piping or ventilated channels[2] are intended as a two-barrier system. Therefore, a failure in the inner pipe, outer pipe or channel must be detectable. A simultaneous failure of inner and outer pipe/channel is not considered. Note that this concept only works if both barriers are monitored![3]

---

[1] The expressions "two-barrier" and "2-barrier" are used here synonym.

[2] The IGF-Code use the term "duct" instead of "channel".

[3] On open deck the ambient is assumed to be a sufficient secondary barrier.

© The Author(s), under exclusive license to Springer Nature Switzerland AG 2025
G. Würsig, *The Safety Principles for the Use of Low Flashpoint Fuels in Shipping*,
Synthesis Lectures on Ocean Systems Engineering,
https://doi.org/10.1007/978-3-031-64174-9_5

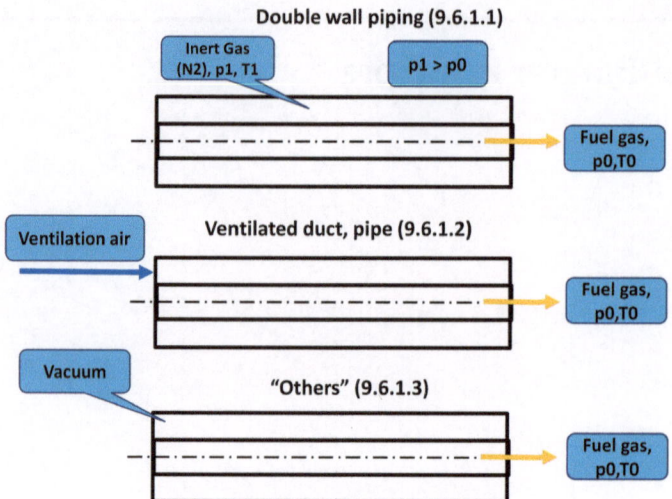

Fig. 5.1 Double barrier principle for pipes and ducts [1]. (*Source* GMW Consultancy)

The regulations related to the double wall piping as given by Sects. 9.5 (fuel distribution), 9.6 (Gas-Save-ER), Sect. 9.7 (ESD engine room (ESD-ER)), Sect. 9.8 (general requirements) are a very good example for prescriptive requirements which "fall a little bit short".

The double barrier principle for double walled piping and ducts is illustrated in Fig. 5.1. For the application in engine rooms comp. Section 9.2, p. 83.

After lengthy discussions during the IGF-Code development if it is thinkable to have any different arrangement than the double wall piping (Sect. 9.6.1.1), it was agreed that there might be different solutions. The result was to allow ventilated ducts (Sects. 9.6.2 and 9.8). After more discussions it was also agreed that the idea to have a controlled double barrier by evacuation of the space between the pipes is a possibility.[4] This solution is named more generic in Sec. 9.6.1.3 by stating "other solutions providing an equivalent safety level may also be accepted by the Administration".[5] This is where the regulation "fall short". For more details comp. the grey box below.

---

[4] This was and may be still is regarded by the "two pressure promotors" (Sect. 9.6.1.1) as a completely crazy one.

[5] Maybe the intention was: we will find a way to exclude these strange ideas.

"Short fall":
Inerting the annular space with Nitrogen is a very good solution for gas in the inner pipe at atmospheric temperature. It starts to become complicated but manageable if the temperature is low as e.g. with LNG in the inner pipe because the Nitrogen pressure varies with changing temperature.

With LH2 as fuel makes it difficult to use Nitrogen because it becomes solid at LH2 boiling temperature related to 1 bar pressure. These systems would need Helium as gas for the annular space. Helium is not only expensive it is also a rare gas with limited supply sources. It should not be wasted for unnecessary applications.

An alternative solution is a moderate vacuum in the annular space. This can be applied in all cases and with all gases. The vacuum can easily can be supplied by a vacuum pump which also can be used to control the system tightness.

May be to exclude ventilated ducts as a practicable solution, the requirement that these ducts have to be pressure resistant to the pressure of the inner pipe (Sect. 9.8.2) was introduced. After subsequent additional lengthy discussions, it was agreed that a practical test might demonstrate the applicability of a large duct which is not designed for the full pressure of the inner pipe.[6] Hopefully this possibility will not disappear from the IGF-Code in future with the reasoning that "it is not needed"!

The second short fall of the Sect. 9.8 and also other regulations in the IGF-Code is the aim to limit the pressure to less than 10 bar. This cannot be technically justified at all and makes it difficult to design systems at a pressure level for direct injection piston engines[7] and gas turbines.[8]

This limitation has again the historical motivation that the first engines were engines with mixture formation before the turbo charger (at approx. 1 bar) and later on with mixture formation just before the cylinder (at approx. 5–8 bar).

With a small safety margin, all pressures above 10 bar have been regarded to be much too dangerous to be used and the 10 bar was defined as "high pressure". To a process engineer who knows that large reactors in chemical industry are safely operated well above 10 bar and even up to pressures above 1.000, −bar, this argument appears strange. For many shipbuilding engineers this is obviously not the case.

---

[6] IMO [1], Sect. 9.8.2 last sentence: "As an alternative to using the peak pressure from the above formula, the peak pressure found from representative tests can be used..."

[7] Design is for approx 300 bar or above.

[8] Needed pressure is well above 30 bar.

5.2   Double Block and Bleed Valves

A Double-Block and Bleed is a valve arrangement with two block and one bleed valve (Double-Block and Bleed) and is also used in land applications but not very often. Land based industry in general accepts a double block valve arrangement as an adequate measure of redundant protection. Figure 5.2 shows an example with the easiest, most illustrative but also most space and money consuming solution with three valves.[9]

The bleed is necessary to depressurize the space between the valves. In general, bleeding is done to the ambient. The bleed stays open to ensure that a leak of one of the valves will create a flow through the bleed but no flow to the downstream side of the valve arrangement. The system can also be used to depressurize the piping. The principle as described above has the disadvantage that fuel gas is released to the ambient and that air can enter the space between the block valves which might require purging before operating the block valves again.

The author's view is that a better solution in most cases is to guide the bleed to a low pressure reservoir,[10] to close the bleed after depressurisation and to control the tightness by a pressure monitoring between the block valves. Such an arrangement ensures that the two barrier principle with verification of the functionality of both barriers is met. This proposal

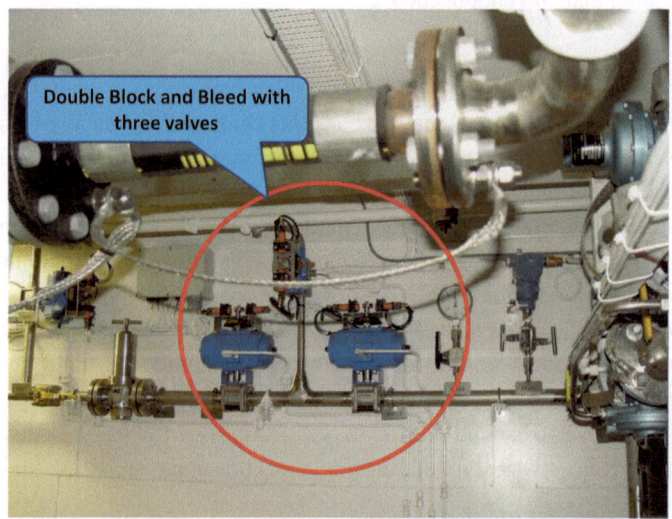

Fig. 5.2 Double block and bleed arrangement with three valves. (*Source* Dr. Gerd Wuersig)

---

[9] Arrangement with two valves (one three way valve and one single way valves) are more cost efficient.

[10] E.g. the fuel tank.

goes beyond the Code requirements because the IGF-Code, IGC-Code do not require the monitoring of the bleed.

5.3   Other Two-Barrier Principle Arrangements

For a safe system engineering it is relevant to apply the two-barrier principle not only to the cases named above but in general. Namely the design principles for type C fuel tanks is an example of "a hidden" application of the two-barrier principle (comp. Sect. 6.5, p. 46). For the other tank types this is more obvious but as shown in Chap. 6 all fuel and cargo tank design principles of IGF-Code, IGC-Code [1, 2] follow the 2-barrier principle.

Two other examples for two-barrier principle design are:

- Single wall fuel piping on open deck. The secondary barrier is the release to ambient in case of any leak. Secondary barrier control is not ensured and not needed in this case.
- Use of an air lock between spaces with possible fuel release if installed below deck. Alternatively, access from open deck to these spaces is possible without an air lock.

References

1. IMO (2016), IMO Resolution MSC.391(95), IGF-Code: International Code of Safety for Ships using Gases or other Low-Flashpoint Fuels, IMO, London, ISBN 978-92-801-1653-3
2. IMO (2016), IMO Resolution MSC.5(48), MSC.370(93), IMO IGC-Code, International Code for the Construction and Equipment of Ships Carrying Liquefied Gases in Bulk, IMO London, ISBN 978-92-801-1631-1

Cargo and Fuel Containment Systems

6

The fuel containment system is a practically unlimited source of burnable media, which in future may be even toxic.[1] All other parts of the fuel system contain only a limited amount of fuel which is much easier to handle because the consequences of a failure are much lower.[2] This section discusses the safety principles used to ensure a separation of the fuel containment system from the rest of the ship and the upstream fuel process system.

6.1 Overview

The design of practically all tanks for liquefied gases according to the IGF-Code and IGC-Code [1, 2] apply the two barrier principle (comp. Chap. 5, p. 29).[3] The aim of the tank regulation for gas carriers is to ensure the survival of the ship in case of a major tank failure which is not related to a collision. The large amount of vapours which would be created in case of a major tank failure are tolerated because the purpose of the ship is to transport the gas as cargo. It is important to note that the liquefied gas tanks for gas fuelled ships are based on the gas carrier requirements and regulations. In fact, the regulations for LNG are nearly identical.

For gas fuelled ships it is difficult to tolerate large gas releases which might endanger crew and passengers on board because the purpose of the ship is to transport goods or passengers.

---

[1] In case of Ammonia as fuel.

[2] Note: bunkering systems are to some extend an exception. The possible release is much smaller compared to the tank content but in most cases higher than in other parts of the fuel system (comp. Chap. 11, p. 91).

[3] Note: IGC-Code Sect. 4.4.2 allows deviations as Sect. 6.4.2.2 of IGF-Code is doing.

© The Author(s), under exclusive license to Springer Nature Switzerland AG 2025   35
G. Würsig, *The Safety Principles for the Use of Low Flashpoint Fuels in Shipping*,
Synthesis Lectures on Ocean Systems Engineering,
https://doi.org/10.1007/978-3-031-64174-9_6

The liquefied gas is only the fuel. The goal for fuel containment systems according to Sect. 6 of the IGF-Code is of major importance:

"6.1 **Goal**"
"The goal of this chapter is to provide that gas storage is adequate so as to minimize the risk to personnel, the ship and the environment to a level that is equivalent to a conventional oil fuelled ship."

In addition the special FR for the section defines:

"6.2.1 the fuel containment system shall be designed that a leak from the tank or its connections does not endanger the ship, persons on board or the environment...."

This principle is valid for LNG as fuel (Part A1 of [1]) anyhow it is the authors opinion that it should be applied for all alternative fuels!

Tanks for liquefied gases are always designed to be liquid and gas tight under normal operation and maintenance conditions. In practice leaks from liquefied gas tanks in normal operation are very rare. However, failure conditions which mainly relate to fatigue are considered for all tank types to ensure the fulfilment of the secondary barrier principle.

How the two barrier principle works for the main tank types for ships with liquefied gases as cargo (IGC-Code, [2]) and as fuel (IGF-Code, [1]) is explained in the following.

The tank types discussed here are defined in the IGF-Code [1]. For ships with alternative fuels the tank types are defined by the following sections of the IGF-Code:

- 6.4.15.1 Type A independent tanks
- 6.4.15.2 Type B independent tanks
- 6.4.15.3 Type C independent tanks
- 6.4.15.3 Membrane tanks
- 6.4.16 Limit sate design for novel tank concepts

The same tank types are defined in the IGC-Code [2] for gas carriers in the following sections:

- 4.21 Type A independent tanks
- 4.22 Type B independent tanks
- 4.23 Type C independent tanks
- 4.24 Membrane tanks
- 4.27 Limit sate design for novel concepts

The following section mainly make reference to the origin of all this which is the IGC-Code [2]. The intention during the development of the IGF-Code [1] was to apply the same design criteria as was applied for gas carriers. Even if the Part-A1 is only related to LNG as fuel, it should be noted that the tank regulations are valid for all liquefied gases transported with gas carriers. Overall these are 37 different liquefied gases regulated in the IGC-Code [2] plus Hydrogen. The later is covered by an *IMO Interim Guideline* [3].

> **Note on LH2 ship transport:**
> The interim Guidelines on the transport of LH2 were initiated by Japan and Australia because Japan wanted to start a pilot project on the import of LH2 from Australia. This project matured and in 2022 the LH2 gas carrier **MS Suiso Frontier** went into service [4]. The LH2 tank of the ship is designed in the conventional way as a vacuum insulated tank with a state of the art design.
>
> Obviously, the project partners (Kawasaki, Shell and others) detected that it is technically not feasible to build large tanks for LH2 gas carriers using this design principle. This was also one of the conclusions from the German and EU *R&D* projects in the 1990*s* [5, 6].
>
> Definitely the Suiso Frontier project is a real milestone because it simply did what has been missing more than 25 years before in Europe. It practically demonstrated that such a transport is possible, initiated the development of the currently not existing technology for this transport and the development of rules for large scale LH2 transport.
>
> Consequently, Japan and Australia proposed the development of revised guidelines for LH2 ship transport which are intended with a finalisation target in 2024 [7, 8].
>
> Compare Sect. 6.6.3, p. 56 for more details.

Two different tank design principle are distinguished. The first are defined as independent tanks according Sect. 4.1.4 of IGC-Code [2]. "...Independent tanks are self-supporting tanks. They do not form part of the ship's hull and are not essential to the hull strength. The tanks are independent from the ship structure....". Type A, B and C tanks are independent tanks.

The second design principle is named membrane tank design. According to Sect. 4.1.5 of IGC-Code [2] "...Membrane tanks are non-self-supporting tanks that consist of a thin liquid and gastight layer (membrane) supported through insulation by the adjacent hull structure....".

Beside these practically dominating tank design principles, the IGF-Code and IGC-Code are open for new developments and provide the possibility to develop new designs by using the limit sate design concept (comp. Sect. "4.27 Limit state design for novel concepts" [2]). The guidelines for such designs are given as Annexes to the IGC-Code and the IGF-Code.[4]

---

[4] IGC-Code: Appendix 5–Standard for the use of Limit State Methodologies in the Design of Cargo Containment Systems of Novel Configuration; IGF-Code: ANNEX: Standard for the use of Limit State Methodologies in the Design of Fuel Containment Systems of Novel Configuration.

To the knowledge of the author so far no tank designs have been approved on the base of the Limit state design requirements.

It should be noted that tanks for gas carriers cover a typical transport capacity range between 10.000, $-m^3$ up to more than 200.000, $-m^3$ per ship. The typical LNG carrier of today has a capacity of approx. 170.000, $-m^3$ which is equal to approx. 1.000, $-$GWh of delivered energy. Nearly all large LNG carriers have membrane tanks. Nevertheless to illustrate the size of a large LNG tank ,as done in Fig. 6.1 the *Moss type tank* may be the better choice because it allows the direct comparison of the sphere of the St Peter's Basilica in Rome (58, 90 m diameter) with the sphere of the inner tank of a 177.000, $-m^3$ Moss carrier built 2019 by MHI in Japan (44, $-$m diameter).

In contrast to these large capacities, the need for gas as ship fuel is much smaller. The largest energy requirements in shipping are related to large container ships. Figure 6.2 give an example.[5]

The LNG fuelled container ship in Fig. 6.2 has a single membrane tank of 18.000, $-m^3$ volume. This is the upper end of the volume range needed for LNG fuelled ships. The energy content is sufficient for a round voyage China/Europe/China. At the other end of the scale of seagoing ships with LNG as fuel are ships which operate in a limited area like the ferry operating in the Baltic Sea shown on Fig. 6.3.

Fig. 6.1 Illustration of LNG tank sizes by comparison of a 44.000, $-m^3$ Moss tank with the size of the St Peter's Basilica in Rome. (*Source* GMW Consultancy)

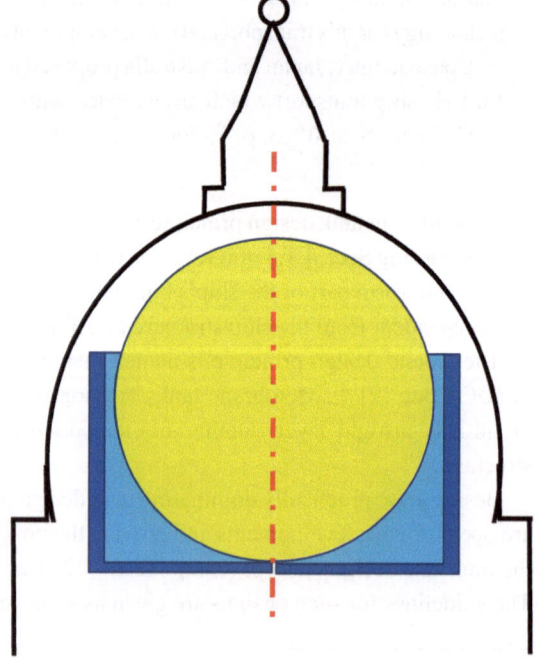

---

[5] The person in the foreground is the author.

Fig. 6.2 The ship in the background is a typical large Container ship: MS Trocadero, $23.000, - TEU$, built 2021. (*Source* Dr. Gerd Wuersig, 2.04.2023)

Source: Dr. Wuersig, 2013

Fig. 6.3 MS Viking Grace in 2013: 2 LNG tanks of 200 m^3 each installed at the aft of the ship. Tanks are type C tanks of Wartsila LNG-Pack design. (*Source* Dr. Gerd Wuersig)

In between these volume range for LNG fuelled ships are large LNG fuelled Cruise ships. Figure 6.4 show the AIDAnova which is the first Cruise ship built for LNG as fuel.

Cruise Ships need an LNG volume of $2.000, - m^3$ to $6.000, - m^3$. The main tank type is type C. Figure 6.12 show such a tank.

In the following, the two barrier principles for Type-A, Type-B, Membrane and Type-C tanks are explained.

Fig. 6.4 Cruise ship AIDAnova during bunkering. First Cruise ship with LNG as fuel. Built in 2018 by Meyer Werft. (*Source* c/o: AIDA Cruises)

> **Note on 2-barrier illustrations**:
> Figures 6.5, 6.7, 6.8 and 6.9 illustrate the 2-barrier principle for the tank types.
> - They are not to scale.
> - They do not contain any shipbuilding design. E.g., all ships have a double bottom which is not illustrated, tank supports are not included, etc.
> - All tank types have insulations in most cases. Type C tanks for Propane, Butane transport at ambient temperature don't have insulations. Insulation is only indicated for the Membrane Tank Fig. 6.8 because for the 2-barrier principle it is only relevant for this tank type.

6.2   Type A Tank

The tank is designed following "...classical ship structural analysis procedures..." (Sect. 4.21.1.1 [2]). The tank pressure is low and limited to 0, 7 bar g. For these tanks it cannot be excluded that a *fatigue crack* develops into a *fatal crack* . The result of such a fatal crack would be a leakage of the entire tank content. To mitigate this risk, a complete secondary barrier is required for a type A tank (Fig. 6.5).

The space between tank the wall and the secondary barrier must be able to contain all tank content which might be released from the tank. It should be noted that a possible evaporation of liquefied gas into the inter-barrier space may lead to a large vapour formation which requires adequate pressure relief from the inter-barrier space (indicated by the red

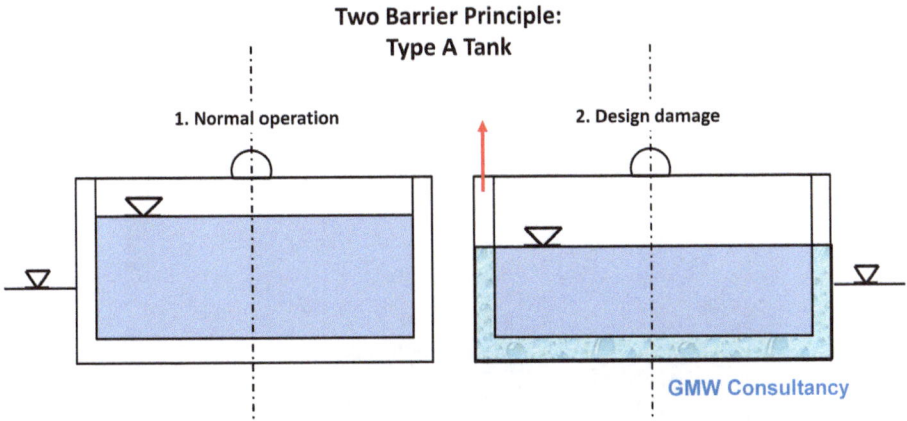

Fig. 6.5 The two barrier principle for Type A Tanks (for IGC-C, IGF-C). (*Source* GMW Consultancy)

arrow in Fig. 6.5). The secondary barrier must be able to withstand possible low temperatures from the liquefied gas release.

For gas carriers with type A tanks according to the IGC-Code the outer shell of the ship can be the secondary barrier. Typical type A tank gas carriers are carriers for LPG. Please note that also other ships with other cargoes than LPG are very often named LPG Carrier. The term is used as a synonym for liquefied gases with a boiling point at 1 bar above −55 °C. The reasoning is very practical. For cargoes above −10 °C any ship hull material can withstand the low temperature of the liquefied gas and down to −55 °C ship grade steels are available for hull design which allow the outer ship shell to be defined as secondary barrier. Below −55 °C the secondary barrier would be from stainless steel or aluminium. This would be too expensive. For this reason, the general Type A gas carrier is following the temperature limit of −55 °C which is in the range of the boiling temperature for propane at 1 bar (−42 °C).

It should be noted that Type A tanks with an independent secondary barrier have been developed and built for LNG as cargo and fuel by LNT Marine. The 2020 built gas carrier SAGA DAWN (IMO 9769855) with 45.000,−m³ LNG capacity is the first LNG carrier with a Type A tank design. Figure 6.6, p. 42 show the tank installation into the ship.

The LNT A BoX is designed as a tank in tank system where the outer tank liquid tight surface of the insulation acts as a secondary barrier preventing any damaging effect to the hold space or the outer shell even in case of any leak from the inner tank. The insulation system is fitted to the hold space, while there is no insulation on the primary tank. Thus, the system features an IMO type A tank without insulation inside an insulated hold space, like a cold box–and thereof the name A-BOX.

Fig. 6.6 LNT Marine Type A Tank ("LNT A-Box") during lifting into the tank hold of SAGA DAWN. (*Source* LNT Marine Pte., Ltd.)

Type A-Tanks for liquefied gas transport are typically used for large LPG Carriers which today have a ship cargo capacity of up to 110.000, $-m^3$ with 4–5 tanks. Approx. 730 out of the approx. 2.300 gas carriers are Type A gas carriers (32%). The Type A design has been developed for LNG carriers by LNT Marine (comp. Fig. 6.6). The first ship with 45.000, $-m^3$ LNG tank capacity went into service in 2019 named SAGA Dawn (IMO No 9769855).

Ammonia as cargo and fuel:
A number of gas carriers with type A tanks are qualified to transport NH_3. In fact all projects related to the large scale PtX transport are assuming that the NH_3 will be transported with Type A tank gas carriers. This might be a practical starting point for the import of PtX NH_3. In the long run, the existing Ammonia carriers are much too small for energy import. They have an Ammonia cargo capacity of approx. 300 GWh. With this they are by far to small for efficient large scale energy import.

As described above (Sect. 6.1, p. 38) the typical LNG carrier has a cargo capacity of approx 1.000, $-GWh$ which is more than 3 times the value the largest NH_3 carriers have. For more details about ship import of PtX the reader may refer to [9] (specially Sect. 2.2.4). On behalf of MAN Energy Solutions SE the author contributed to the part of the project related to PtX import by ship and pipeline.

It should be obvious from Fig. 6.5 that the failure scenario includes the venting of large amounts of gas created by the boiling of liquefied gas in the cargo hold. In fact the pressure relief of type A tank gas carriers for this case are large flaps on deck.

At least from the author's point of view, the use of type A tanks in future for NH_3 fuelled ship applications does not appear to be a good idea. Note that the related fuel tanks would have to be more than two times bigger than an LNG tank with the same energy content .

6.3   Type B Tank

Type B independent tanks are designed to ensure that a *fatigue crack* in the inner shell do not develop towards a *fatal crack* without additional load changes. The design ensure that the leak rate from such a tank is limited and predictable. Of course, a leaking type B tank cannot be operated any more without repair.

Section 4.22 of the IGC-Code requires "... model tests, refined analytical tools and analysis methods to determine stress levels, fatigue life and crack propagation characteristics....". The model testing in particular makes the development of type B tanks expensive. The benefit of this tank type is that it only requires a partial secondary barrier which is able to handle the possible leak rate from the tank. Figure 6.7 illustrates the principle for a spherical tank of *Moss design* and a prismatic tank of *SPB design*. For prismatic type B tanks the pressure is limited as for type A tanks to 0, 70 bar g.[6]

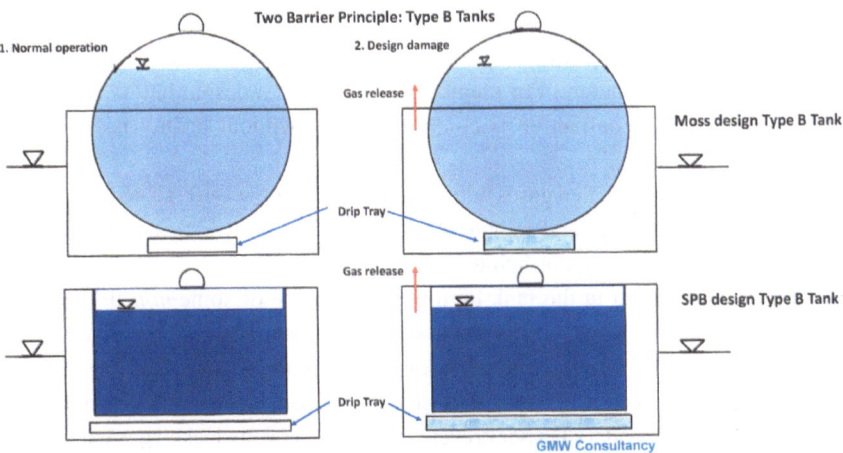

Fig. 6.7 The two barrier principle for Type B Tanks in IGC-Code, IGF-Code. (*Source* GMW Consultancy)

---

[6] Comp. Sect. 4.22.1.1 IGC-Code, Sect. 6.4.15.2.1.1 IGF-Code.

The regulations for Limit state design developed during the revision of the IGC-Code between 2008 to 2010 according Sect. 4.27 of IGC-Code [2], Sect. 6.4.16 of IGF-Code [1] are based on the regulations for type B tanks.

The typical Type B-Tank is a *Moss design* tank and is used for LNG tankers. Tank material is mostly Aluminium. The typical size is $25.000, -m^3$ to $30.000, -m^3$. Latest Moss type B tank carriers have a tank size of up to $44.000, -m^3$. The ships have a typical cargo capacity of to $160.000, -m^3$. Ships with up to $180.000, -m^3$ have been built. Nowadays nearly all LNG cargo tanks are Membrane Tanks and not Type B Moss tanks. E.g., 53 LNG carriers above $50.000, -m^3$ cargo capacity went into service in 2021. Only one of them was a Moss type LNG carrier [10]. Approx 18% or 110 out of the approx. 600 LNG carriers which were in operation in 2022 have type B tanks. The total gas carrier fleet are approx. 2.300 ships.

Some gas carriers with prismatic *SPB design* tanks have been built (less than 10). Tank material is mainly also Aluminium. Most ships have been built for LPG transport. Ship cargo capacity for these ships is typically below $100.000, -m^3$.

For LNG fuelled ships the share between membrane and Type B prismatic tank is approx. $50/50$ for tank sizes above $6.000, -m^3$.

6.4   Membrane Tank

Membrane type tanks are generally not independent from the ship's hull. They are an integrated part of the ship structure. For membrane tanks the "....design basis...is that thermal and other expansion or contraction is compensated for without undue risk of losing the tightness of the membrane."[7]

Membrane tanks for liquefied gases with a temperature below $-10\,°C$[8] require a complete secondary barrier because a fatal fatigue crack is used as the worst case design criterion. Figure 6.8 illustrates the design principle.

The liquid is contained in the tank by a thin membrane of some *mm* thickness which is supported by supports integrated in the first layer of insulation and the insulation itself. Practically the weight of the liquefied gas is distributed into the insulation.

A possible leak in the first membrane is guided by channels through the first insulation and vented. The thin secondary barrier prevents the contact of any leaked gas with the second insulation. The second insulation is supported directly by the ship's inner hull.

---

[7] Section 4.24.1.1 of IGC-Code, 6.4.15.4.1.1 of IGF-Code.

[8] Note that this temperature level is used for type B and membrane tanks because in principle the outer ship's hull may be used as secondary barrier for higher gas temperatures even if low grade steel is used.

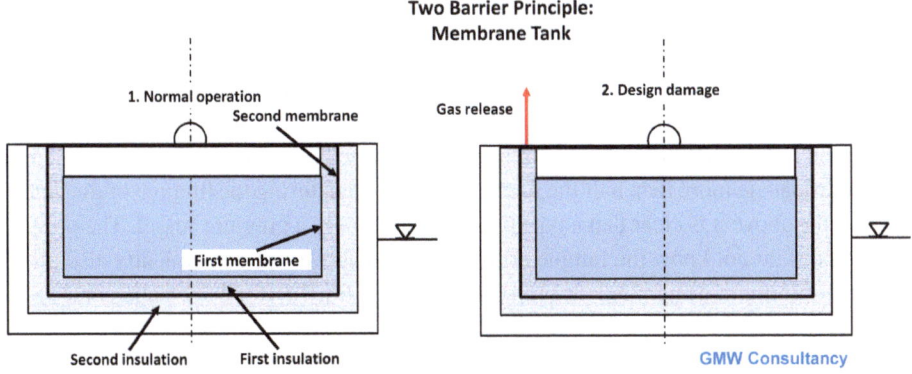

Fig. 6.8 The two barrier principle for Membarne Tanks (for IGC-C, IGF-C). (*Source* GMW Consultancy)

The second insulation prevents the hull from reaching to low temperatures even in case of occurrence of the design failure. As for the other tank systems the total insulation thickness is about 300, −mm to 400, −mm. A special design feature is that the integration into the ship structure leads to the need to design a fixed bearing and a movable bearing in the horizontal direction.

The typical LNG carrier today has Membrane Tanks. Approx. 80% out of 600 vessels in 2022 (total gas carrier fleet is approx. 2.300, − ships). The size of the ships goes beyond 250.000, −m^3 with 5 tanks. However, the standard Membrane Tank LNG carrier today has 170.000, −m^3 to 175.000, −m^3 with most times 4 tanks. These ships deliver approx. 1 TWh of energy. For the complete supply chain obviously this capacity has been proven to be commercially competitive to pipeline transport and oil as energy carriers.

Also, about 50% of large ships with LNG as fuel use membrane tanks with a maximum capacity of approx. 18.000, −m^3 for a single tank (comp. p. 38). The other ships with such large fuel tanks use type B tanks.

Practically all membrane tanks for LNG ship transport and LNG fuel have been designed by GTT (Gaztransport & Technigaz, 1, route de Versailles, 78470 Saint-Remy-les-Chevreuse, France). The yards built the systems based on licence agreements with GTT.

For the reader who may ask himself if it is a good idea to use membrane type tanks for NH_3 as fuel? Even if the amount of possible gas release is much less than for type A tanks, the authors opinion is the same as for type A tanks for the same reason (comp. Sect. 6.2, p. 43).

6.5   Type C Tank

Type C independent tanks are designed "...on pressure vessel criteria modified to include fracture mechanics and crack propagation criteria. The minimum design pressure defined...is intended to ensure that the dynamic stress is sufficiently low, so that an initial surface flaw will not propagate more than half the thickness of the shell during the lifetime of the tank."[9]

From the above it is clear that a type C tank is **NOT** only a pressure vessel. The pressure vessel Codes do not know the fatigue criterion named above and consequently also do not apply the minimum design vapour pressure according Eq. 6.1 (for IGC-Code comp. Sect. 4.23.1.2)!

$$P_0 = 0,2 + A \cdot C \cdot \rho_r^{1,5} \tag{6.1}$$

The factor A is related to the membrane stress in the tank shell, the factor C to the tank dimensions and the ρ_r is a dimensionless density factor related to the highest cargo or fuel density. At the end, the formula ensure that any type C tank has a minimum wall thickness independent from the design pressure related to the process design.

The result of the above is the assumption that the design of a type C tank exclude any leak caused by fatigue! Therefore, no additional barrier is required. The secondary barrier is a built in design feature as illustrated by Fig. 6.9.

For fatigue crack propagation calculations, the load cycle is most relevant. In general, the most relevant load cycles are the cycle of pressurisation/depressurisation and the thermal cycle of cooling down/warming up. When the type C concept was developed the tanks had a size of approx. $1.000, -\text{m}^3$ and practically only cylindrical shape (comp. below). Assuming a short distance voyage for such a type C gas carrier may lead to one loading and one

Fig. 6.9 The two barrier principle for Type C Tanks (for IGC-C, IGF-C). (*Source* GMW Consultancy)

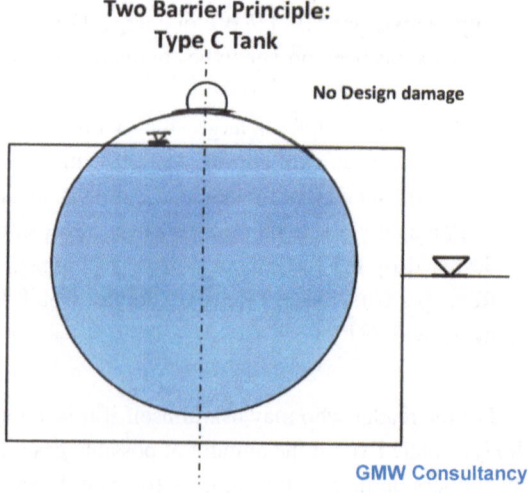

Two Barrier Principle: Type C Tank

No Design damage

GMW Consultancy

---

[9] Sections 4.23.1.1 of [2], 6.4.15.3.1.1 [1].

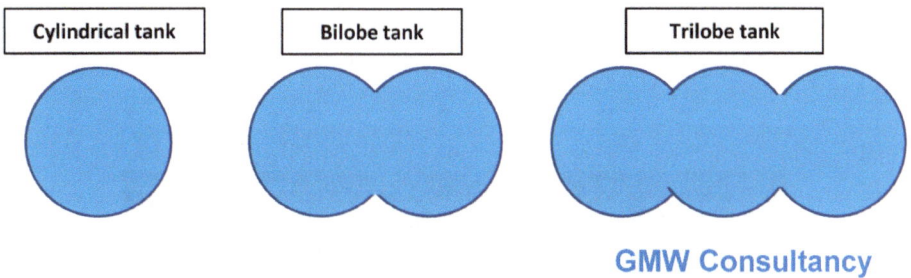

Fig. 6.10 Different shapes of type C tanks. (*Source* GMW Consultancy)

unloading per week. This may give 52 load cycles per year. Assuming a 25 year design life result in 1.300, −load cycles. In addition it can be concluded from the regulations that the wave loads have been also considered.[10]

The design principle works very well in practice. The author is not an expert in fatigue life calculation and therefore cannot judge if the one-to-one transfer of the design basis mentioned above to gas as ship fuel regulations already consider that a fuel tank is subject to much more and faster pressure changes than it has been assumed in development of the type C tank calculation base.

Type C tanks are not only built with a pure cylindrical shape. Also bilobe tanks and trilobe tanks are built for a better space utilisation. Figure 6.10 illustrates the principle shape.

Figure 6.11 show a type C tank of a Cruise ship. The tank is one of three tanks which have a combined volume of approx. 4.000, −m³ LNG.

For LNG fuelled ships type C tanks also have been built in a vertical configuration. Such an arrangement might be favourable for LH2 because it has the potential to reduce the heat flux through the supports (comp. also Sect. 6.6.3, p. 56). Figure 6.12 give an example of such a tank.

Figure 6.13 shows the interior of one half of a large Bilobe type C tank. The pump tower and bottom filling line are similar in all tank types. The leader and the two platforms may illustrate the tank size.

> Type C Tank gas carriers started with small tank sizes of approx. 1.000, −m³ to 3.000, −m³. Today the most common Type C tanks have approx. 5.000, −m³ to 10.000, −m³. The ships have cargo capacities of 25.000, −m³ to 60.000, −m³. Designs for nearly 100.000, −m³ exist. The typical tank pressure is between 4, 5 bar g to 7, 0 bar g.

---

[10] Last sentence IGC-Code 4.23.1.2 "When a specified design life of the tank is longer than 10^8 wave encounters..." the allowable dynamic stress factor "...shall be modified to give equivalent crack propagation corresponding to the design life."

LNG fuelled ships with tank capacities below approx. 4.000, $-m^3$ are nearly completely equipped with Type C-Tanks. Single tank size may be 2.000, $-m^3$ and above.

Fig. 6.11 Type C tank for a Cruise ship built by Meyer Werft. (*Source* c/o: AIDA Cruises)

Fig. 6.12 Vertical Type C tank of approx. 1.000, $-m^3$ tank volume. (*Source* TGE Marine Gas Engineering GmbH, Bonn, Germany)

Fig. 6.13 One half of a Bilobe Type C tank of approx. 15.000, $-m^3$ bilobe tank volume. (*Source* TGE Marine Gas Engineering GmbH, Bonn, Germany)

For the reader who may ask himself if it is a good idea to use type C tanks for NH_3 as fuel: the authors opinion is that they are by far the most safe solution if Ammonia is used as fuel because a failure of the tank shell is much more unlikely compared to type A, B and membrane tank type (comp. also Sect. 6.2, p. 40)

6.6 Others

In this sections some left overs related to tank system saftey and future hydrogen transport are discussed.

- The separation principle for fuel tanks tank connections to fuel tanks are explained.[11]
- The principles to ensure equivalent safety between portable fuel tanks and fixed installed fuel tanks are presented.[12]
- An outlook to the possible development of PtX production is given[13] as a motivation to overcome in the near future the technical challenge to transport LH2.[14]

6.6.1   Tank Connections to Fuel Tanks

This section explains the relevance of tank connections as the weak point for the release of large amounts of fuel into the ship. The illustrating example is related to type C tanks which have been the first tank type used for LNG as ship fuel. The explained process technology is also valid for the other tank types.

On gas carriers all piping is concentrated in a tank dome located on the top of the tank. No piping is penetrating the tank shell below the maximum liquid level. The safety reason for this is that any damage to a pipe from the tank and before the first valve in the pipe will lead to a flow from the tank until the liquid level in the tank is below the pipe penetration into the tank. Considering the dimensions of the tanks, this principle for pipe connections is regarded to be important.

With the first LNG fuel storage tank systems this principle was not followed any more. This was probably because the designers simply did a "copy and paste" adaption from land applications of small, pressurized tanks for LNG road transport and cryogenic liquid storage in industrial applications. These tanks have very often have a bottom line for loading and unloading, and also a bottom connection for a pressure built up heat exchanger. The tanks are vacuum perlite insulated and have ferritic steel outer shell which will get brittle fractures if cooled down to cryogenic temperatures.

The process technical reason for this is that the cryogenic pumps need a positive suction pressure above the boiling pressure of the liquefied gas in the tank and that the pressure built up heat exchanger need also a pressure difference between the gas space and the flow to the heat exchanger to enable a flow without an additional pump. This is simply the transfer of the technical solutions from shore to ships. It needs to be considered that the well established onshore design needs to be modified at least considering the following differences to shore applications:

- Piping on board ships are subject to movements and vibration loads which do not occur on shore applications.

---

[11] Section 6.6.1, p. 50.

[12] Section 6.6.2, p. 53.

[13] Section 6.6.3, p. 56.

[14] Section 6.6.4, p. 58.

- The possibilities to mitigate a large release of liquefied gas or liquid low flashpoint fuel into a ship are much more difficult on board of ships compared to shore applications. E.g., the ship compartments are closed to the bottom of the ship and need to be watertight.
- The escape from board of a ship is something different compared to the escape from a shore facility. It is much more difficult to press the shutdown button for the complete plant (ship) operation, and to run away than it is on shore site plants.

Probably related to the approval by class society departments who also approve gas carriers, and perhaps more likely because the engineers in these departments who contributed to the development of the IGC-Code in the 1970s were still on the job, the simple use of shore designs got not an approval. The tanks were modified to reach a level of safety as known from type C tanks without pipe connections below the liquid level. For this reason, these tanks where equipped with stainless steel outer shells which do not have a risk of brittle fracture. Additionally, the "tank connection space" was created which is defined in the IGF-Code as follows:

"2.15.3 Tank connection space is a space surrounding all tank connections and tank valves that is required for tanks with such connections in enclosed spaces."

In other words, it was intended to cover the vacuum space of the LNG tank up to the first valves in the piping from the tank. For this space the two barrier principle was applied again by requiring a stainless steel vacuum jacket and a valve chamber for the first valves which could be flooded with LNG without dangerous effects to the rest of the ship. Figure 6.14 illustrates the principle. For a summary of the different space types for alternatives fuels, please comp. also Chap. 10, p. 85.

For LNG and other alternative fuels for ships it is often necessary to install the process equipment to load and unload the tank below the liquid level of the tank. This is a principle difference to gas carriers where all this equipment is installed above the tanks, or at least at the level of the tank dome. With process equipment below the tank liquid level, it is much more difficult to stop a leak from the tank even if the piping is installed on the tank top. Not only the pressure difference between tank and ambient but also a possible syphon-effect leading to flow in case of pressure equilibrium between ambient and tank gas space should be mitigated.

In addition, the hold space was required to be designed from material able to withstand the low temperatures in the hold space in case the vacuum space is flooded with LNG. For this scenario, the safe routing and handling of cold fuel gas from the vacuum space must be ensured.

If no connection below the liquid line is used, the heat exchanger might be placed below the minimum liquid level to allow a liquid flow at low tank filling levels. Figure 6.15 illustrates the principle.

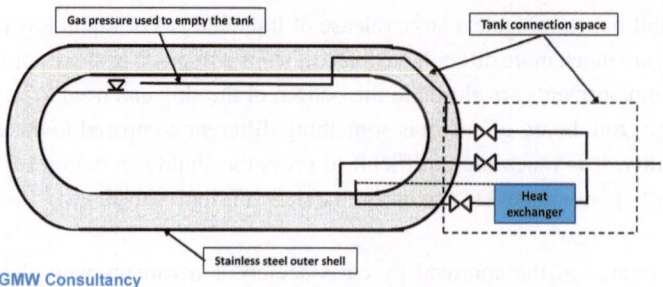

Fig. 6.14 Type C vacuum insulated tank with bottom connection and without double walled piping (early design: tank connection space include insulation space; stainless steel outer shell). (*Source* GMW Consultancy)

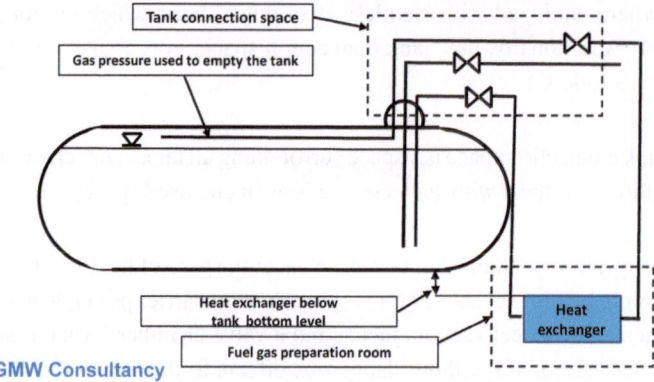

Fig. 6.15 Type C tank without connections below liquid level. Conventional insulation or vacuum insulation with conventional steel outer shell. Double wall piping outside process spaces are not illustrated. (*Source* GMW Consultancy)

In the design as illustrated in Fig. 6.15 the heat exchanger is not placed in the tank connection space. It can be placed anywhere but preferable below the lowest liquid level of the tank to ensure a flow with low pressure difference between boiling pressure of the liquid and gas pressure in the gas space. Higher locations are also possible. In a initial start phase of the pressure built up the needed pressure might be provided by Nitrogen which can be mixed to the fuel during normal operation. Note that a possible syphon-effect should be considered (see above).

Between 2009 and 2010 Tarbit, Wartsila and Germanischer Lloyd did the first application of the Wartsila LNGPac system. The chemical tanker "Bit Viking" was the first ship with an LNGPac and the first LNG fuelled ship classed with Germanischer Lloyd (Fig. 6.16).

The outer shell of the vacuum insulated LNG tanks are built with normal ferritic carbon steel. The fuel tank has a bottom line which is enclosed by a second stainless steel pipe acting as the secondary barrier. Nowadays such designs are generally accepted.

Fig. 6.16 Application of the first LNG-Pac system from Wartsila on Germanischer Lloyd classed chemical tanker Bit Vicking (owned by Tarbit). (*Source* Wartsila)

With today's calculation procedures it should be also possible to qualify a bottom line according to the requirements outlined for type C tanks in Sect. 6.5. Such a qualification would have the result that no double wall piping of the bottom line would be necessary.

6.6.2 Portable Fuel Tanks

An important part of the regulations on LNG as ship fuel and also for other alternative fuels is the regulation on the equivalence of portable fuel containment systems and fixed installed fuel containment systems. The motivation for this regulation (Sect. 6.5 IGF-Code [1]) arose from the observation that designers proposed the use of standard containers for LNG as fuel very soon after the (IGF-IG) had been published in 2009 [11]. A reasoning for this proposal was that the bunkering of fixed installed tanks in port was not established and bunker ships were not available. Going back to implemented standard portable ISO containers was a solution to get things running. To avoid the costly and "unnecessary" installation of fixed tanks may have been also a motivation in some cases.

It was very soon common sense among administrations and class societies that such an approach would compromise safety, but, at the same time, would be very attractive because it would be much cheaper than to follow all the "completely unnecessary" requirements for fixed tank installation. On the other hand, it was and is obvious that it should be possible to use portable tank systems.

During the development of the IGF-Code, France proposed the regulations now included as Sect. 6.5 in the IGF-Code. That these regulations were needed and appropriate became obvious to the author when the discussion of the complete text of this section of the code at the working group's meeting in London only lasted about one hour. This was very fast for a working group which liked to consider every word and comma twice. Main items of the regulation are:

- Only type C tank designs are allowed.[15]

  - Remember that a type C tank is not only a pressure vessel (comp. Sect. 6.5, p. 46) and of course it is not a standard container for liquefied gas.

- The tanks must be placed in dedicated areas and not "'somewhere".[16]
- Mechanical protection against cargo operation failures, spill protection, water spray systems must be provided on open deck and, for below deck storage, the space is regarded as a tank connection space (comp. also Sect. 6.6.1, p. 50).[17]

  - It is not permitted to place some LNG containers "somewhere" on a Container ship and to use them as fuel tanks.[18]

- The tanks must be appropriately secured to the deck.[19]

  - It is not permitted to use only ordinary twist locks.

- Ship stability must be observed.[20]
- Connections must be made using approved hoses and piping.[21]

  - The use of industrial standard connections without an approval for gas fuelled ships is not permitted.

- Spill from inadvertent disconnection or rupture of non permanent connections must be covered by adequate technical arrangements.[22]

---

[15] IGF-Code, Sect. 6.5.1.

[16] IGF-Code, Sect. 6.5.2.

[17] IGF-Code, Sect. 6.5.2.1.

[18] Note that the IMO IMDG-Code [12] for containerised transport does not permit the use of the cargo in these containers during transport!

[19] IGF-Code, Sect. 6.5.3.

[20] IGF-Code, Sect. 6.5.4.

[21] IGF-Code, Sect. 6.5.5.

[22] IGF-Code, Sect. 6.5.6.

- The pressure relief system must be connected to a fixed venting system.[23]

  – It is not permitted to blow gas from a PRV into the container bay.

- Control, monitoring and safety systems must be integrated into the ship's control, monitoring and safety systems.[24]

  – It is not permitted to have a local temperature and pressure measurement only.

- Safe access to tank connections must be ensured.[25]

  – It is not permitted to install the system in the middle of a container bay, five rows high and connect them without a possibility to access the connections.

- Each portable tank in a multi tank installation must be capable to be isolated at any time.[26]

  – It is not permitted to connect, e.g. 10 tanks, with hoses and without any shut down possibility.

- The *FL* for fixed tanks apply.[27]

  – It is not permitted to connect e.g. 10 tanks and to have a 5% vapour space on the top tank and fill the rest to 100%.

In addition, the requirements for bunkering (IGC-Code Sect. 8.) apply to portable tanks re-filled on board.

The author is persuaded that a number of approval engineers in class societies have burnt a candle in church, temple or wherever because with these regulations endless discussions came to an end. It also was a load off the minds of experienced designers and yard managers to have a clear guidance. Hopefully the headache does not come back with the introduction of other alternative fuels and the storage tanks for CO_2 related to CC on board ships.

---

[23] IGF-Code, Sect. 6.5.7.

[24] IGF-Code, Sect. 6.5.8.

[25] IGF-Code, Sect. 6.5.9.

[26] IGF-Code, Sects. 6.5.10.1 and 6.5.10.2.

[27] IGF-Code, Sect. 6.5.10.3.

6.6.3 Upcoming PtX Production

A number if not the majority of experts believe that LH2 transport will never be relevant for PtX import by ship. The author opinion is different. On the long run[28] LH2 will be the cheapest solution for PtX import by ship (comp. Fig. 6.17).

At the end of 2023 the UN world climate conference in Dubai[29] declared the aim to triple the "renewable energy" capacity until 2030.[30] Figure 6.18 show the historical and predicted energy need of the world until 2100 (green line) and the historical and predicted growth in world population (black line).

All figures for 1965–2022 are real figures. Consumption figures mainly based on [13], population figures based on UN population statistics. The dotted blue line is the predicted fossil share if the share of CO_2-free energy supply remains at the same percentage as it had been in 2018.

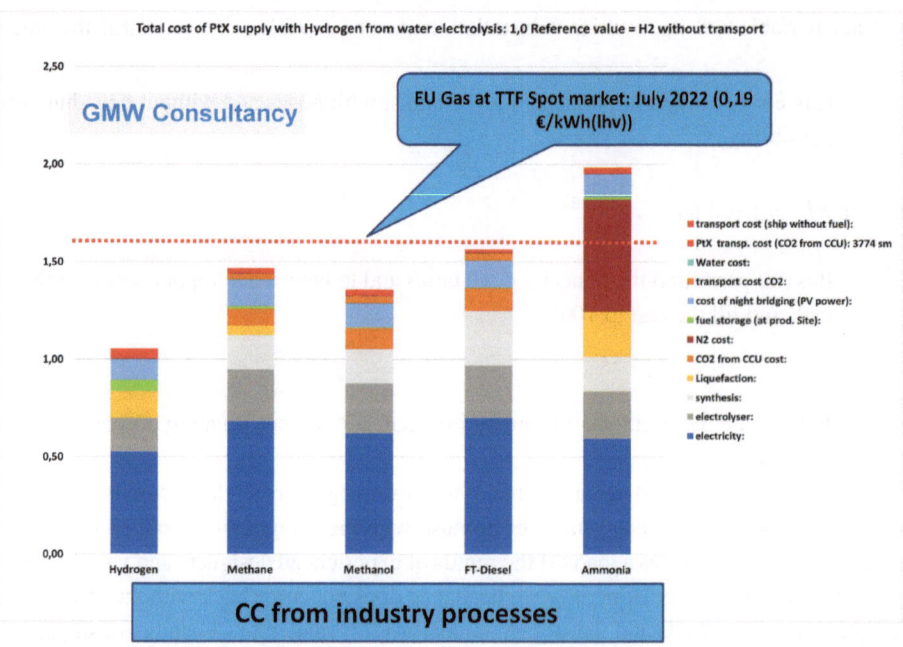

Fig. 6.17 Relative Power to X production costs (1, 0 is equal to Hydrogen production costs without transport). (*Source* GMW Consultancy)

---

[28] Beyond 2040–2045.

[29] Dubai, 30 November to 12 December 2023.

[30] Closing Document COP-28: item 28.
"28. Further recognizes the need for deep, rapid and sustainable reduction in greenhouse gas emissions in line with the 1,5 °C pathway and calls on Parties to contribute to the following global efforts…Tripling renewable energy capacity globally …. by 2030 ….".

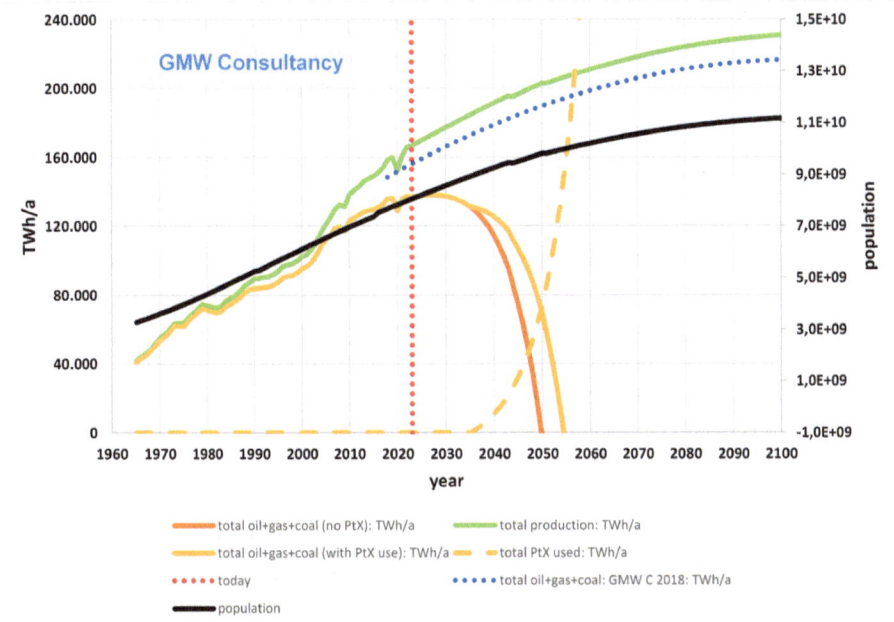

Fig. 6.18 Result of threefold the 2022 wind and solar power in 2030 and keeping the growth rate of CO_2 free energy. History and prediction of world energy need and world population growth based on GMW C prediction model 2019. (*Source* GMW Consultancy)

The solid orange and yellow lines indicate the fossil energy need if the growth rate for CO_2 free energy[31] supply is achieved and remains at the same percentage level beyond 2030.[32]

As a most likely scenario, it is assumed that energy transport from production locations to consumption palaces will be needed as it is today with fossil fuels. Prior to the export of CO_2 free energy, the need at the production places will be covered. Therefore, it is assumed that approx. from 2030 onward an export of CO_2 free energy is needed because for a large number of countries the local consumption will be to low compared to the production.[33]

In addition, it is assumed that the production of CO_2 free energy is not sufficient to allow a pipeline export only and that the needed pipelines will not be available in time. For these reasons, a PtX export will be needed. It is assumed that it will reach relevant export volumes from 2035 onward.

From the above scenario it is concluded that shipping and air traffic will be able to use PtX fuels to a relevant extend from the second half of the next decade onward. The easiest supply

---

[31] Solar, wind, nuclear, hydro power, bio mass.

[32] 100% always the year before.

[33] But not for Europe, US, China. At least if these areas can keep their industrial basis.

will be for Asia/Europe Container trade because only two bunkering places are needed for this trade. These are Rotterdam and Singapore. These ports are relatively close to potential PtX production areas in Africa respectively Australia. Considering that a ship has a 25 year design life lead to the conclusion that the time for preparation is now.

The dashed yellow line in Fig. 6.18 indicates the total used PtX and the solid yellow line indicates the resulting decrease in fossil energy production. Please note that the additional effort related to the fact that more electricity is needed for PtX production compared to direct use of electricity is considered. This results in a difference between orange and the yellow line because for the orange solid line no PtX production is assumed. In other words, the assumption for the solid orange line is that any energy produced by renewable sources can be used close to production without the need of long distance transport.

If the anticipated growth in CO_2 free energy supply become reality the real figures will be somewhere between the solid orange and yellow lines.

Please note that the author's opinion is that the scenario for CO_2 free energy supply as indicated by assuming the growth figures as requested in the COP-28 document are too ambitious. Assuming half of that growth rate seems more realistic to the author. This would shift the zero fossil use of the orange and yellow line from 2050, 2055 toward 2074, 2083. Considering this, the author assumes that PtX energy supply will start with Methane (CH_4), FT Diesel like fuels, Methanol (CH_3OH) and Ammonia (NH_3) with relevant amounts approx. from 2035 onwards.

Over time these energy carriers will be substitute by Liquefied Hydrogen (LH2) and Hydrogen transported by pipelines. Such a scenario would fit into historical timelines for changes in energy carrier from wood to coal, from coal to oil and ongoing shift from oil to natural gas.

Figures 6.17 and 6.18 show results form PtX price and world energy need modelling which have been developed by the author since 2019. The energy need values given by Fig. 6.18 until 2022 are mainly based on data from BP [13]. The forecast has been developed by the author in 2019 based on BP data until 2018. It is unchanged and, as the reader can see, matches the real figures at least until today.[34]

6.6.4   Hydrogen Transport and Heat Flux into the Tank

For the introduction of LH2 as an energy carrier it need to be considered that LNG carriers can not be used for LH2 transport.

The following example may explain why existing LNG carriers are not suitable for LH2 transport and why the simple scaling of the existing technology is also not a practicable option.

---

[34] Of course the values from 2018 to 2022 have been changed to real consumption figures. The further predictions are not changed since 2019 modelling.

For liquefied gas tanks on gas carriers and gas fuelled ships the heat flux into the tanks is generally expressed by the Boil Off Rate in %/d at 100 % filling (BOR). The gas is called Boil Off Gas (BOG). The BOR is a common way to express the heat flux from the ambient into the liquefied gas tank. It is defined by:

$$\dot{Q}_{fg} = \dot{M} \cdot (h'' - h') \ [W] \tag{6.2}$$

$$\dot{q}_{fg} = \dot{Q}_{fg}/A_{Ta} \ [W/m^2] \tag{6.3}$$

Typical BOR rates for LNG tanks are $0, 12$–$0, 20\%/d$ for large tanks of LNG carriers and $0, 3\%/d$ to $0, 60\%/d$ for LNG fuel tanks. For the Moss tank as illustrated in Fig. 6.1 the application of Eqs. 6.2 and 6.3 is given in Annex B, p. 133. The same calculation is also done for LH2 (comp. Annex B, p. 133).

Assuming that the problem of condensing Nitrogen and Oxygen from air at the outer tank wall of the LH2 tank is solved and all tank materials are suitable for LH2, the result is that the insulation thickness must be increase from $400, -$mm to $5.400, -$mm which is a factor of $13, 5$.

The simple increase in insulation thickness is only a part of the solution because the heat transfer through the bearings alone would lead to a BOR of still $0, 9\%/d$.

From the calculations given in Annex B, p. 133 it is concluded that a transport of LH2 with existing LNG carriers is not possible and a simple adaption of existing LNG designs is not the way forward. The Japanese experiences with the LH2 carrier Susio Frontier confirm this judgement. The BOR of the Susio Frontier is more than $1, 3\%/day$ which confirm the calculation presented in in Annex B, p. 133.[35] Regardless of this, there are more clever ways to design a LH2 carrier tank system with reasonable BOR.

With the background explained above, Japan started an initiative at IMO to develop special rules for new LH2 tank designs. The tanks are under development in Japan as a result of the Australian/Japanese LH2 project (comp. above on page 35). This development in Japan and at the IMO is promising for the progress of shipping liquefied hydrogen and disappointing for the author because looking back to his own work some decades ago, it looks like the reinvention of the wheel (comp. Sect. 6.1, p. 35).

References

1. IMO (2016), IMO Resolution MSC.391(95), IGF-Code: International Code of Safety for Ships using Gases or other Low-Flashpoint Fuels, IMO, London, ISBN 978-92-801-1653-3
2. IMO (2016), IMO Resolution MSC.5(48), MSC.370(93), IMO IGC-Code, International Code for the Construction and Equipment of Ships Carrying Liquefied Gases in Bulk, IMO London, ISBN 978-92-801-1631-1

---

[35] BOR Susio Frontier calculated by the author from published technical data.

3. IMO RESOLUTION MSC.420(97) (adopted on 25 2016), INTERIM RECOMMENDATIONS FOR CARRIAGE OF LIQUEFIED HYDROGEN IN BULK, IMO, London 2016
4. Weltweit erster Wasserstofftanker beladen, Wirtschaftswoche, 21. Januar 2022
5. Dr.-Ing. G. Würsig, Seetransport und Verwendung von Wasserstoff: Forschung und Entwicklung 1986 bis heute, Vortrag: DKV-Jahrestagung, Würzburg 20.11.98
6. G. Wuersig, Shipping Liquid Hydrogen, Marine Engineers Review, December 1991, p. 10 to 14, London 1991
7. IMO Submission MSC 104/15/11, Australia, Japan, 2 July 2021; Revision of the interim recommendations for carriage of liquefied hydrogen in bulk; IMO London, 2021
8. IMO submission CCC 8/14, Australia, Japan, 14 2022; REVISION OF THE INTERIM RECOMMENDATIONS FOR CARRIAGE OF LIQUEFIED HYDROGEN IN BULK, Proposals for revision of the structure of the Interim Recommendations, and the way forward; IMO, London 2022
9. Project Report 2023; GreenH2SZ-Conceptualization of a feasible green hydrogen supply for the Salzgitter region within the context of the European hydrogen strategy-; Wasserstoff Campus Salzgitter; John-F.-Kennedy-Str. 43–53, 38228 Salzgitter; MAN Energy Solutions SE, Fraunhofer-Institut für Schicht- und Oberflächentechnik IST; free download: http://wasserstoff-campus-salzgitter.de/projektbericht-greenh2sz/
10. The LNG industry GIIGNL Annual Report 2022; GIIGNL-International Group of Liquefied Natural Gas Importers, 8, rue de l'Hôtel de Ville-92200 Neuilly-sur-Seine-France
11. IMO RESOLUTION MSC.285(86)(adopted on 1 2009); INTERIM GUIDELINES ON SAFETY FOR NATURAL GAS-FUELLED ENGINE INSTALLATIONS IN SHIPS; IMO, London, 2009
12. INTERNATIONAL MARITIME DANGEROUS GOODS CODE (IMDG-Code); IMO London
13. bp Statistical Review of World Energy 2022, 71st edition, BP PLC, London

Safety Valves for Liquefied Gas as Cargo and Fuel Tanks

7

The fuel tanks of gas fuelled ships and gas carriers are protected by safety valves which are intended to protect the tanks against unacceptable tank pressure. For this reason they are called PRVs.[1] Often the sizing principles and with this the adequacy of the pressure relief system are not known and consequently not understood. This may lead to dangerous misinterpretations because the tank PRV is most relevant for tank protection.

This chapter aims to throw some light onto the background of sizing, need, and risks of these PRVs. As explained below the sizing principles are not only related to shipping they are also very closely related to land based installations in process- and refinery industry.

As well as the explanation of the background of sizing principles, the risks related to the pressure relief and possible mitigations are highlighted.

7.1 PRV Sizing

As a general principle the PRV should protect the tank against overpressure which might damage the tank. For this reason, the PRV system should be based on the largest possible source of pressure increase for the tank. This may include the sizing with respect to the maximum flow related to tank filling. In other words, the maximum filling or bunker rate.

For the design and use of safe systems it is necessary that the designer, approval authority and user is aware of the relation between the risk of overfilling and PRV sizing even

---

[1] Comp. IGF-Code Sect. 6.7, IGC-Code Chap. 8.

© The Author(s), under exclusive license to Springer Nature Switzerland AG 2025   61
G. Würsig, *The Safety Principles for the Use of Low Flashpoint Fuels in Shipping*,
Synthesis Lectures on Ocean Systems Engineering,
https://doi.org/10.1007/978-3-031-64174-9_7

if the pressure relief system could or is not designed to release the full capacity of loading/bunkering. In the following, the explanation of PRV sizing is limited to the standard case according Sect. 6.7.3 of the IGF-Code and Sect. 8.4 of the IGC-Code.

7.1.1 Sizing Against External Fires

The standard sizing case is related to the protection of the tank against an external fire. The principle is related to the application of liquefied gas and substances with similar behaviour. The design is based on the assessment of fire risk mitigation in refinery and land transport.

The sizing requirements are all based on the same origin which are the requirements of the American Petroleum Institute (USA Institution developing saftey standards in refinery industry) (API) and Compressed Gas Association (USA Institution developing saftey standards for transport and use of flammable gases) (CGA). Even if this is not known anymore by most experts, the formulas in the relevant Codes are identical[2] as it is shown in the calculation given in Annex C, p. 137.[3] Annex C, p. 137 include the mathematical relationship between API, CGA, IGC-Code and IGF-Code and is not repeated in the main part of this book.

It should be recognised that the different heat fluxes caused by a fire of 71 kW/m^2 (IGC-Code, IGF-Code), 109 kW/m^2 (CGA), 66 kW/m^2; 43 kW/m^2 (API) are "the same" and that the difference is related to different mathematical calculation procedures and safety factors (comp. Annex C, p. 137). The basic figure derived from experiments with hydrocarbon fires for a tank fully engulfed by the fire is always 109 kW/m^2. The area used is the tank surface. This figure is a realistic figure for relatively small tanks as e.g. liquefied gas tanks on trucks which are involved in a large road accident. Larger tanks in refineries or on ships will not be completely engulfed by a fire.

7.1.2 The Fire Case

The determination of the size of PRVs for gas carriers and natural gas fuelled ships are sized based on the assumption of fire which is partly engulfing the cargo/fuel tank. The basics related to the development of the formula used in the Codes can be found in the publication by Heller [1] and in Appendix C, p. 137. The origin of this goes back to research work done in the 1950s of last century. As basic figure the maximum heat flux of a hydrocarbon fire engulfing a liquid filled tank was used for dimensioning

$$\dot{q} = 34.500, - BTU(h \cdot ft^2) \cdot \approx 109 \text{ kW/m}^2 \qquad (7.1)$$

---

[2] Not only similar.

[3] Comp. Sect. 1.1 for reasoning why this is not known anymore.

The requirements consider that a large tank is not completely engulfed by fire by application of the factor 0, 82 to the tank surface. This factor is used to determine the total heat flux into the tank. By doing this the heat flux from Eq. 7.1 can be defined by Eq. 7.2 to

$$\dot{q} = 70.961, - W/m^2 \approx 71 \text{ kW}/m^2 \tag{7.2}$$

This is the basic figure used in IGC-Code and IGF-Code for calculating the "...vapours generated under fire exposure...".[4]

This figure is somewhat hidden in the Codes and not quoted in the Codes. To hide the origin a little bit better the formula in the Codes are naming the result Q. This is very close to the $\dot{Q}$, usually used for a heat flux which has the unit kW. In fact the result is a heat flux but converted to a volume flow of air at standard conditions of 1 bar, 0 °C and named "...minimum required rate of discharge of air at standard conditions of 273, 15 Kelvin (K) and 0, 1013 MPa...".[5] The possible reason is very practical because the unit can be used directly to choose a valve from the catalogue of a manufacturer which are using this unit to define the valve capacity.[6]

In fact it is confusing to someone who is trying to evaluate what energy input is used in aiming to calculate the related evaporation rate of cargo/fuel which is necessary at least to calculate the pressure drop in the vent line up- and downstream of the PRV as it is required, e.g., by IMO-Assembly Resolution A.829(19), 23. Nov. 1995 [2] and explained in [3].

As the pressure drop for, e.g., Methane gas is definitely not the same as for air IMO A.829(19) [2] require to calculate with the properties of, e.g., Methane and NOT air. However, the IGF-Code [4] is calculating with air for "Sizing of vent pipe systems".[7]

This has its origin in a proposal prepared for IMO by an Administration at the final stage of IGF-Code development. Prior to the decision on this subject in 2012/2013 the author tried to clarify this. But in vain. The argument that the author has developed the original regulation [2] together with colleagues from Det Norske Veritas (Classification Society based in Norway) (DNV) and Society of Gas Tankers and Terminal Operators (SIGTTO) did not help. This mistake is therefore still a part of the IGF-Code. For those who like to know why this should not be the case please comp. Sect. 12.1.5, p. 108.

The heat flux into a tank partly engulfed by a fire can be calculated using the specific heat flux $\dot{q}$ according Eq. 7.2 and the reduced tank surface area as explained above:

$$\dot{Q} = \dot{q} \cdot A^{0.82} \approx 71 \cdot A^{0.82}; [kW] \tag{7.3}$$

Note that the result is equivalent to the calculation of the "...vapours generated under fire exposure..." in IGF-Code, IGC-Code (comp. above). The benefit of Eq. 7.3 is that it

---

[4] IGF-Code Sect. 6.7.3.1.1.2, IGC-Code Sect. 8.4.1.2.

[5] Same number in IGF-Code, IGC-Code as above.

[6] It may be that it is the other way round and the valve manufacturers give this value because the rules use it.

[7] IGF-Code Sects. 6.7.3.1.3 and 6.7.3.2.

allows the physical correct interpretation of the so called "...fire exposure factors..." used in the Codes to consider the protection of the tank from a fire. This invention is the only invention done for shipping application. At least the value of the fire factors "F" are unique for shipping. Refinery industry and gas industry use different reduction factors in their regulations (comp. Annex C, 137).

With the fire factors Eq. 7.3 become:

$$\dot{Q} = F \cdot \dot{q} \cdot A^{0,82} \approx F \cdot 71 \cdot A^{0,82}; [kW] \tag{7.4}$$

Equation 7.4 is equivalent to the equation used in IGC-Code Sect. 8.4.1.2 and IGF-Code Sect. 6.7.3.1.1.2

For those who not believe or simply want to check, comp. Annex C, p. 137, GASTECH-1994 contribution [1, 3]. This will also answer the question why 71 and not $109^{0,82}$ is used (see Eq. 4 in Annex C, p. 137).

7.2　PRVs Sizes for LNG Fuelled Ships

In this subsection the fire factors for the sizing of the tank PRVs, the size of the PRVs and the need for a maximum working pressure for LNG fuel tanks are discussed. At the end of the section some recommendations for the PRVs on gas fuelled ships are given.

7.2.1　Fire Factors for PRV Sizing

Both Codes use the same fire factors.[8] $F = 1, 0$ for uninsulated, $F = 0, 5$ for insulated tanks on deck and uninsulated tanks in holds, $F = 0, 2$ for insulated tanks in holds and $F = 0, 1$ for membrane tanks and insulated tanks in inerted holds.[9] Note that "uninsulated" tanks in the IGF-Code Part A-1[10] only make sense for PRV sizing of compressed natural gas. For LNG at approx. $-160\,°C$ it makes no sense because any LNG tank must be insulated! On the other hand the sizing procedure assumes a heat flux into a liquid in the tank. Therefore the use for compressed gas would be a misinterpretation.

For vacuum insulated tanks which practically can be built up to a maximum volume of approx. $1.000, -m^3$ some developers of the IGF-Code became innovative and introduced fire factors different from those in the IGC-Code (Sect. 6.7.3.1.2 [4]). They are related to "...vacuum insulated tanks in fuel storage hold spaces and for tanks in fuel storage hold spaces separated from potential fire loads...". Obviously someone experienced a case with a hold space close to high fire loads. It is remarkable that the special fire factors are not related

---

[8] Comp. Eq. 7.3, Sect. 7.1.2.

[9] Definitions here are from IGF-Code but IGC-Code is "the same".

[10] Part for natural gas.

to vacuum insulated tanks on deck. Anyhow the factors valid according Sect. 6.7.3.1.2 are $F = 0, 25$ instead of $F = 0, 5$ and $F = 0, 1$ instead of $F = 0, 2$.

To know the origin of the fire factors which include different mitigation measures for tank protection from the direct fire load would be useful. The origin of the values for the fire factors is hidden, at least for the author, in the darkness of the history of the committee work on liquefied gas transport with gas carriers. To the memory of the author even his admired colleague Martin Böckenhauer, who was a member of the committee for IGC-Code development, had no real explanation for the origin of the values for the fire factors F.

7.2.2 How Big are Cargo, Fuel Tank PRVs?

Figure 7.1 illustrates the fire loads and related valve sizes by a generic example. For surface and volume values used a quadratic shape of the tank is assumed.

Figure 7.1 is related to gas fuelled ships. Therefore, the maximum tank volume is limited to 20.000, $-m^3$ (comp. p. 19, p. 39 for reasoning). Up to 4.000, $-m^3$ to 6.000, $-m^3$ the type C tank is the dominating fuel tank for LNG fuelled ships. On tankers and bulkers these tanks are often placed above deck to limit the cargo loss.

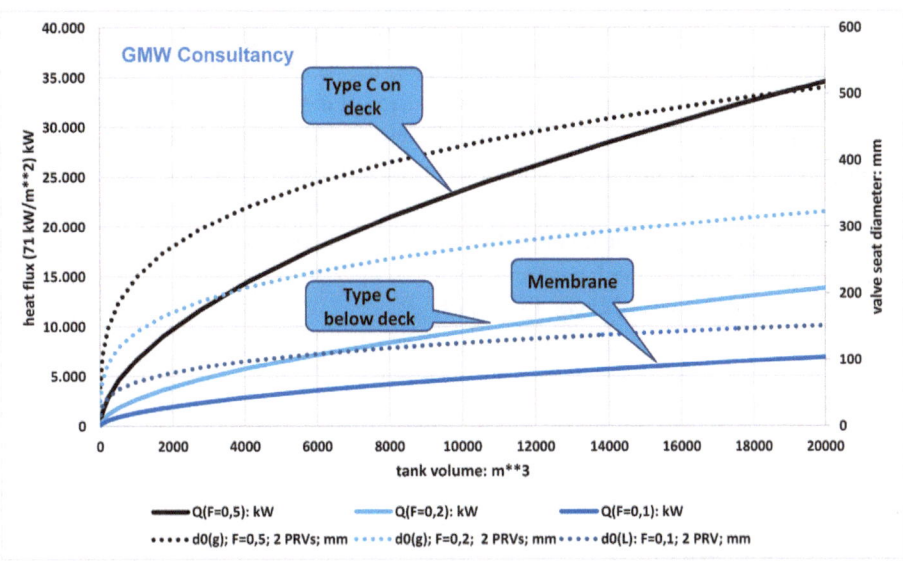

Fig. 7.1 Generic example for calculated energy for the fire case and the resulting PRV seat diameter. Two PRVs per tank are assumed. (*Source* GMW Consultancy)

For these tanks the required fire factor is $F = 0, 5$.[11] The related calculated heat flux from the generic case used for Fig. 7.1 is approx. 9–15 MW.[12] This corresponds to a mass flux through the two PRVs of 18 kg/s to 29 kg/s of Methane. The PRV seat diameter is approx. 260–330 mm.[13]

On general cargo vessels, ferries and cruise ships the tanks are generally installed in a tank hold below deck ($F = 0, 2$). Assuming a typical tank (mostly type C) of 2.000, $-m^3$ gives a heat flux of approx. $4\ MW$, a mass flow of 7, 5 kg/s related to a seat size of approx. 170 mm (two PRVs). For membrane tanks ($F = 0, 1$) between 6.000, $-m^3$ to 18.000, $-m^3$ the values are approx. 4,0–6, 5 MW, 7, 5 kg/s to 13kg/s and 163–220 mm.

At this point the sizing formulas for gas fuelled ships might become dangerous. At the end they lead to large PRVs which, on ordinary cargo vessels, ferries, cruise ships may in itself present a risk. A valve failure would lead to large release of fuel gases. Long vent lines[14] from the tanks, which are normally deep down in the ship, might create additional problems. In case of Ammonia the released gases will not even burnable but also toxic. A careful review and reconsideration of the sizing of safety valves for gas fuelled ships would be very useful. This should include an evaluation of real fire loads for the tank in case of cargo fires.

For gas carriers the story is to some extent different. First, the adequacy of the PRV size has been "tested" (see below). Second, the fuel for the fire will be the cargo and the additional fire load from operating PRVs will not be relevant.

In case of a severe fire the cargo tanks might be really partly engulfed by a fire. In practice the sizing has been found to be sufficient to protect the tanks from failure. The case of the Yuyo Maru is one of the most prominent examples [5]. It has been a kind of "role model" in the discussion about defining the fire factors of the IGC-Code. At that time it was permitted to transport burnable liquids in the side tanks of the ship which is not the case anymore.[15] The LPG carrier Yoyu Maru No.10 had Naphta in the side tanks which ignited when the Pacific Ares caused a rectangular side collision. The ship burned from 9th until 28th of Nov. 1974 when it was sunk by the Japanese Navy.

Fire loads are discussed in Chap. 13, p. 113.

---

[11] Insulated tank on deck.

[12] Single tank volume: 2.000, $-m^3$ to 4.000, $-m^3$.

[13] Note: always 2 PRVs are installed on a tank.

[14] That the pressure losses are always considered carefully seems to the author to be more a hope than a fact.

[15] As often happens, safety learning came the hard way.

7.2.3 Number of Tank PRVs and the Reason for MAWP

The IGC-Code and IGF-Code require two PRVs for cargo/fuel tanks. To "...allow sequential lifting, minimizing unnecessary release of vapour..."[16] or in other words to limit the risk related to the large PRV capacity it is permitted to set one out of the two PRVs to a set pressure up to 5% above Maximum Allowable Relief Valve Setting (MARVS).

This very useful regulation is currently under discussion which seems somewhat strange to the author. One argument is that the flow capacity is limited when only one valve open at MARVS. Considering the fact that the complete valve capacity is reached for the fire case and at a pressure of $1, 2 \cdot MARVS$ this argument is simply not valid[17]. It also ignores that the PRV system is oversized for most practical cases and that a measure of mitigation is needed to limit the full capacity to cases where it is justified.

Finally, a special feature which distinguishes the IGF-Code from the IGC-Code should be mentioned with respect to the PRV system. The IGC-Code only knows one limiting pressure which is MARVS. At this pressure the PRVs open and they open instantaneously[18] and NOT like a proportional safety valve related to a pressure increase above MARVS!

This is also problematic on gas carriers because in practice the cargo system is very often operated very close to MARVS. Experience shows that this seems to be manageable for the transport of cargo according IGC-Code, where the tanks are not part of a dynamically operated fuel system as it is the case for gas fuelled ships. For this reason, Germany proposed for the IGF-Code to introduce the Maximum Allowable Working Pressure (MAWP) concept which is well known and well established in Refinery and Chemical industries.

The IGF-Code require an MAWP with a maximum value of 90% of MARVS[19]. It took 2 years and about 3 working group meetings in London to persuade the group that a limitation of the working pressure below the set point of the safety valves (not only the tank PRVs) is needed.

7.3 Recommendations for PRVs for Liquefied Gas as Ship Fuel

From the above the following principles for fuel tank PRVs are recommended by the author:

- The sizing principles have been applied successfully over decades and should be kept. However, a science based review of the sizing factors F is recommended (comp. Sect. 7.2.2, p. 65).

---

[16] Section 8.2.3 [6], Section 6.7.2.4 [4].

[17] A very much liked misunderstanding is sizing to $1, 1 \cdot MARVS$ because "we always do this". This is true, but only for pressure vessels not designed for the fire case!

[18] "...Full lift safety valves more or less suddenly reach the degree of lift necessary for the mass flow to be diverted following response within a pressure rise of 5%...", Sect. 3.1.2 AD-2000 [7]

[19] IGF-Code, [4], Sect. 6.3.2

Any modifications should be made very carefully. The sizing background should be clearly defined and reflected in any modification. Any review of PRV sizing requirements should be based on a clear and reproducible understanding of the physics. Please comp. Sect. 12.1.5, p. 108.

- The distinction between MAWP[20] as the pressure limit for normal operation and MARVS as the pressure when a PRV starts to open should be carefully observed and controlled.
- PRVs should have a linear opening characteristic to avoid an unnecessary large release. The sizing for the fire case is related to a full valve capacity at "120%" of MARVS **not** to 110% as it is for general applications in storage tank and other pressure equipment design! The full capacity of tank PRVs should be available at 120% MARVS.
- The PRVs should have different sizes to avoid an unnecessary release in cases where the full capacity is not needed. It can be assumed that these cases are the majority of cases. For this reason, the two valves should have different capacities. A 40 and 60% share is regarded to be a reasonable approach. A possible arrangement might be:

  - 40% valve opening at MARVS with full capacity at 105% MARVS.
  - 60% valve opening at 105% MARVS with full capacity at 120% MARVS.

Note that this position was a minority position when the IGC-Code was revised. The current IGC-Code [6] requires PRVs "...of equal size...".[21] As outlined above this might be acceptable on gas carriers.[22]

- It should be noted that safety relief valves are allowed to start leaking tank content at 90% of MARVS.[23] A combined system of a rupture disk and a safety valve (PRV) is a suitable option to ensure valve operation and to avoid any leakage before MARVS is reached. For toxic media like Ammonia this should definitely be a requirement.
- The filling level of the tanks should be limited to ensure that PRVs are not acting because of, e.g., sloshing effects in the tank (comp Chap. 12, p. 99, ff.).
- The standard solution for the filling limit of liquefied gas tanks should be as it is now given in IGF-Code, IGC-Code. Any deviation should be based on an evaluation of operating conditions and possible liquid expansion behaviour. At least for liquefied gas as ship fuel, the maximum FL should be fixed at 98% (comp Chap. 12, p. 99 ff.).

---

[20] Max value 90% of MARVS.

[21] IGC-Code, Sect. 8.2.1.

[22] Comp. Sect. 7.2.2, p. 65.

[23] Remember there are no tight systems in the world (comp. Sect. 3.3, p. 20).

The "Cow Elsa problem"

The interest in pressure relief valves on gas carriers and gas fuelled ships is for the author a "cow Elsa" problem of his professional life. But what is a "cow Elsa" problem?

For the author the "cow Elsa" problem goes back to a sketch of the Comedian Didi Hallervorden (born 1935) presented in German TV (ARD) in 1977. For Wikipedia it goes back to 1115 $p.Chr.$ (Diciplina clericalis, Petrus Alfonsi).

The plot is simple: A servant at a country estate has to inform the landlord about bad news. He wake him up in the middle of the night by phone (it was 1977) and explained that on the landlord's farm the "cow Elsa" has died.

The landlord is not very interested because he has more than a 1.000,- heads cattle on his farm. Anyhow he is now awake and so ask how did "cow Elsa" die? In answer he was told that a beam falling from above killed her.

"Where does the beam come from?" "From the roof of the barn in which cow Elsa was housed." "Why did the beam fall down from the roof?" "Because the barn had burned down." "Why did the barn burn down?" At this point the landlord become more interested. "This was because of the spark coming from the burning manor house." "Why the manor house was burning?" "Because the maid let the candle candelabra fall." "Why you used candle when we have electricity?" "We thought that we should be respectful during the laying out of your wife" (no exact citation).

Since the author has watched this sketch, for him problems with apparently small implications turning into major problems are "cow Elsa" problems.

In the early 90's Dr Krapp (the authors very respected supervisor for about 20 years) came into the author's office and explained that there is a small problem at IMO with PRVs need to be used in 2-phase flow and that he is the one who should take care of this because he did something with 2-phase flow at university.

The author immediately suspected that this was a "cow Elsa" problem. It turned out that it was THE "cow Elsa" problem of his professional life.

References

1. Frank J. Heller, Safety Relief Valve Sizing: API Versus CGA Requirements Plus A New Concept For Tank Cars, American Petroleum Institute 1983 Proceedings-Refinery Department, Vol. 6-2, API, W.DC, p. 123–135
2. Guidelines for the Evaluation of the Adequacy of Type-C Tank Vent Systems, IMO Assembly Resolution A.829(19), adopted on 23rd November 1995, IMO, London
3. M. Böckenhauer, G.M. Würsig (GL), R.H. Chadburn (SIGTTO), B.O. Bauer-Nilsen (DNV),The New Cargo Tank Loading Limit Requirements in the IMO Gas Carrier Codes, GASTECH-1994 Conference proceedings, Kuala Lumpur, 1994

4. IMO (2016), IMO Resolution MSC.391(95), IGF-Code: International Code of Safety for Ships using Gases or other Low-Flashpoint Fuels, IMO, London, ISBN 978-92-801-1653-3
5. Report on the Outline of Collision Between Japanese Tanker YUYO MARU NO.10 and Liberian Freighter Pacific Ares (Nov. 1974), Maritime Safety Agency, Japanese Government, March 1975
6. IMO (2016), IMO Resolution MSC.5(48), MSC.370(93), IMO IGC-Code, International Code for the Construction and Equipment of Ships Carrying Liquefied Gases in Bulk, IMO London, ISBN 978-92-801-1631-1
7. Safety devices against excess pressure-Safety valves -, AD 2000-Merkblatt, Verband der TÜV e. V., Friedrichstraße 136, 10117 Berlin, April 2015.

Collision Protection

As with all modes of transport, accidents related to the movement of the transport vehicle cannot be excluded. Technical measures and regulations aim to reduce the frequency of occurrence and the related consequences, in other words the "risk" (comp. Chap. 2), to an acceptable level. The collision and grounding protection requirements in the IGF-Code aim to ensure that the risk for crew and passengers in such an event is not increased compared to ships fuelled with oil based fuels.

Acceptance is always related to the level of tolerance by society. A touchable example for this effect is the fatality rate in road transport (Fig. 8.1).

In Germany the risk of dying in road traffic[1] was 1970 approx. $2, 7 \cdot 10^{-04}$. At that time this was regarded to be definitely too high, especially because the rate was nearly 20% lower 10 years before.[2] In 1990 the level was reduced by the factor of 2 to approx. $1, 4 \cdot 10^{-04}$ which was nearly the level of 1950 but with $43, 6 \, Mio$ cars compared to $2, 4 \, Mio$ in 1950. In 2020 the risk was at a level reduced by the factor of more than 8 to $3, 3 \cdot 10^{-05}$. From this figure it will be difficult to get further major reductions (see also Chap. 2 for reasoning). Even if lobby organisations and politicians promote the slogan that every "death in traffic is one death to much", it will not be reduced to zero as long as there is traffic. However, this risk of life is accepted by public.

---

[1] Fatalities over inhabitants; before 1990 Bundesrepublik Deutschland and DDR together.
[2] Comp. Chap. 2 for reasoning why a value of 10^{-4} is often regarded to be too high for a technical risk.

© The Author(s), under exclusive license to Springer Nature Switzerland AG 2025   71
G. Würsig, *The Safety Principles for the Use of Low Flashpoint Fuels in Shipping*,
Synthesis Lectures on Ocean Systems Engineering,
https://doi.org/10.1007/978-3-031-64174-9_8

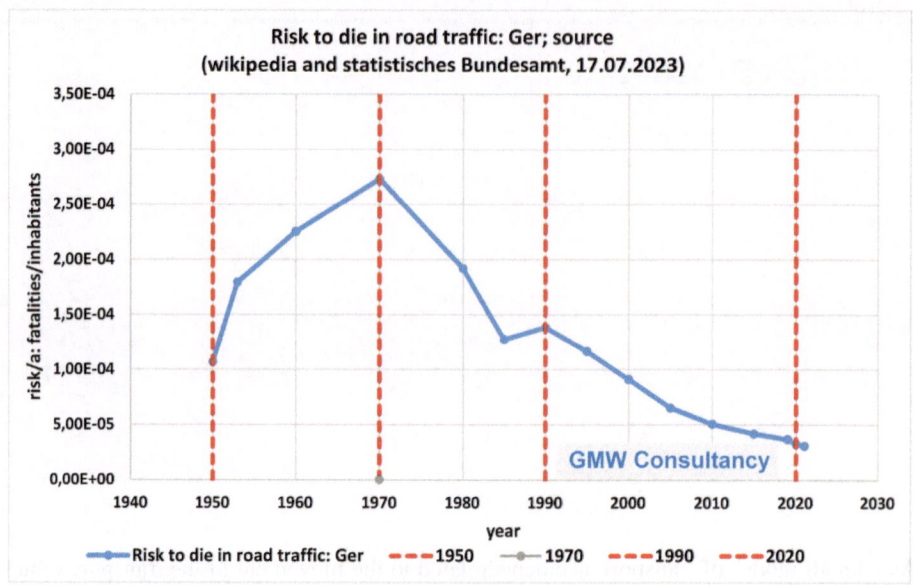

Fig. 8.1 Example for risk acceptance. Fatalities in road traffic in Germany (before 1990: Bundesrepublik Deutschland plus DDR). (*Source* GMW Consultancy)

In shipping thirty "...years ago, the global fleet was losing 200+ vessels a year. At the end of 2022 fewer than 40 losses were reported..." [1], p. 4. "Globally, most of the 27.477, − incidents reported over the past decade have been caused by machinery damage/failure (10.753,−), followed by collision (3.098,−) and wrecked/stranded (2.936)..." [1], p. 4.

As shown in the Table 8.1, the risk of collision for seagoing ships is within the same range as that for grounding and is relatively high (approx. $3 \cdot 10^{-3}$). The risk of death of crew in such an event is below $1, 0 \cdot 10^{-5}$ which is smaller but in the same range as for the road traffic illustrated by Fig. 8.1.

The principles to achieve the aim of an equivalent level of safety for LNG fuelled ships are outlined in this section. The two different requirements for a prescriptive approach with fixed distances defined in Sect. 5.3.3 of IGF-Code and the probabilistic approach considering

Table 8.1 Risk for collision and grounding and for fatalities of crew members (50% yearly occupancy assumed). Figures based on data from [1–4]

	Risk: events/(ship/a)	Risk: death/(year crew)
Collisions	$3, 117E − 03$	$5, 89E − 06$
Grounding (wrecked/stranded)	$2, 853E − 03$	$5, 39E − 06$

the three dimensional location of the fuel tank in the ship, as given in Sect. 5.3.4 IGF-Code are compared by use of a generic example. The example illustrates the position of the author that the probabilistic approach is more adequate to achieve safety equivalence.

8.1 Distance of Fuel Storage as Safety Principle

For gas carriers the protection of the ship from collision, grounding effects to preclude the cargo from release is regulated in Chap. 2., IGC-Code.[3] The IGF-Code has very similar requirements in Chap. 5.3.

The aim of the requirements for gas carriers is to "...ensure that the cargo tanks are in a protective location in the event of minor hull damage, and that the ship can survive the assumed flooding conditions."[4] The goal for ships with LNG gas as fuel is given in Sect. 5.1 of IGF-Code to "...provide for safe location, space arrangement and mechanical protection of...fuel storage systems...". Because the requirements for LNG cargo tanks [5] and fuel tanks [6] are the same, it seems a logical interpretation that both Codes aim to protect against minor collision and grounding events.

In general, both Codes provide the same fixed values for collision protection distances[5] and grounding protection distances.[6] As an alternative to the fixed values for collision distances in 5.3.3.1, the IGF-Code also permits the application of distances based on the likelihood of damage occurrence in case of a collision.[7] This alternative is given in Sect. 5.3.4 of IGF-Code.

Both Codes do not provide criteria for structural energy absorption in case of a collision. Only the distances and therefore the usual energy absorption of a common ship structure is considered. In all cases the LNG tanks must be located behind the collision bulkhead of the ship's forecastle.

Collisions can be grouped into bow, aft, side collisions and groundings. In the following only the side collision and the dependence on tank distance from the side shell is illustrated to explain the principle of tank location for collision protection. Please note that the following explanations explain the principles. They cannot be used directly for design purposes.

---

[3] In the following only requirements for Type 2G and 2PG according to Sect. 2.1.2 of the IGC-Code [5] are considered.

[4] Goal of Chap. 2 [5].

[5] IGC-Code 2.4.1, IGF-Code 5.3.3.1–5.3.3.4 and 5.3.3.6–5.3.3.8.

[6] IGC-Code 2.3.1.2.3, IGF-Code 5.3.3.5.

[7] Probabilistic approach.

8.2   Side Collision Protection

In case of a collision the ship side is effected by the counterpart of the collision. The ramming of an other ship by the gas carrier or a gas fuelled ship is covered by the requirement to locate the tanks behind the collision bow bulkhead. This is not discussed further here.

The IGC-Code and IGF-Code require a minimum distance from the ship's hull. This minimum distance is also applicable for the ship's bottom and is therefore also the minimum space available in case of grounding. Figure 8.2 illustrate the different definitions according to the IGF-Code.

IGF-Code Sect. 5.3.3.1 defines the minimum side distances of the fuel tanks with fixed values for ship breadth (B) divided by 5 or fixed value of 11, 5 m, whichever is less. This is equal to the regulation for gas carriers for the transport of most dangerous gases like Chlorine.[8] This is a very strong regulation for small ships with relatively small fuel tanks.

The regulations to reduce the risk of tank damage in case of collision are mainly related to the breadth of the ship. For the generic examples given in the following a simple breadth, length and fuel storage volume correlation is used as illustrated by Fig. 8.3.[9] The graph (1) gives the tank volume related to the breadth of the ship (left y-axis) and the graph (2) gives the used length of the ship related to the ship's breadth. For all cases it is assumed that the tank is installed at least partly below the waterline as it is common for all cases where the tank is not installed on the weatherdeck.

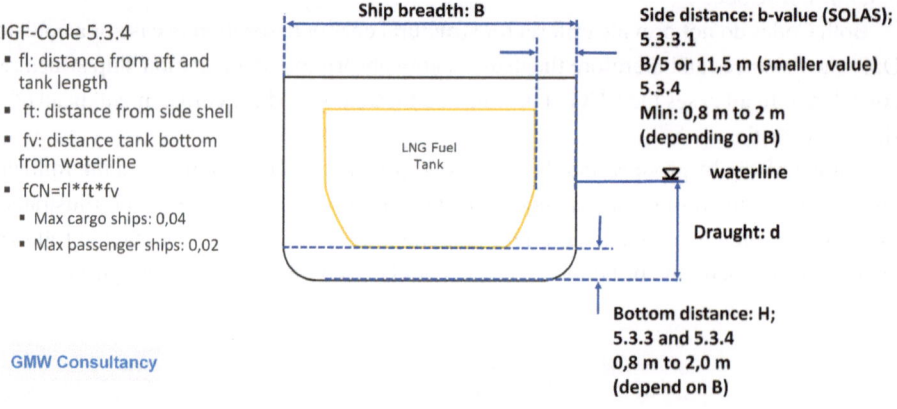

Fig. 8.2 Distances used in IGF-Code [6]. (*Source* GMW Consultancy)

---

[8] IGC-Code Sect. 2.3.1.1.2, for G1 type gas carriers: out of 37 liquefied gases these are: Chlorine, Ethylene Oxide, Methyl Bromide, Sulphur Dioxide.

[9] An energy equivalent LH2 tank will have 2, 4 times and an liquefied Ammonia (LNH3) tank will have 1, 6 times the volume given in Fig. 8.3.

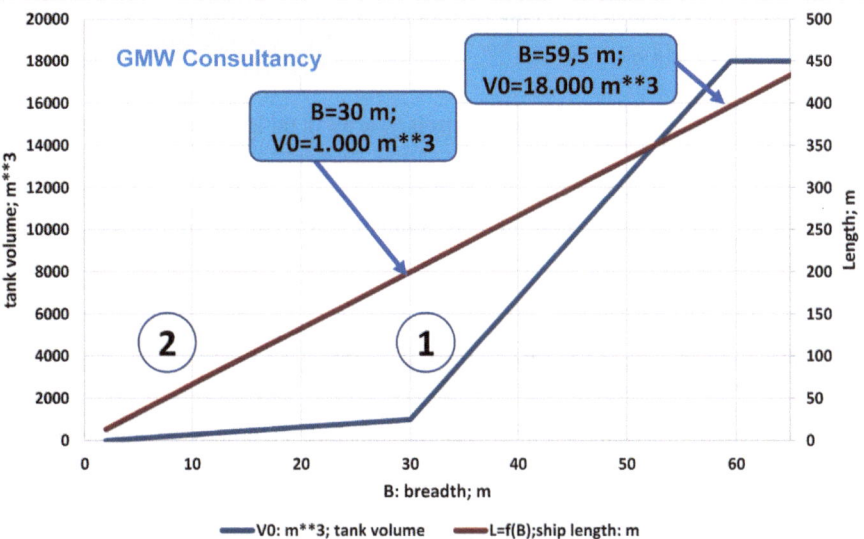

Fig. 8.3 Relations between length, tank volume and breadth used for the generic example cases in this section. (*Source* GMW Consultancy)

Nearly all smaller ships with LNG as fuel use Type C tanks. For Fig. 8.3 it is assumed that Type C are used up to 1.000, −m³ tank volume. Above this size prismatic membrane, SPB and A-Type tanks are assumed. Note that in most practical cases Type C tanks are used up to 4.000, −m³. The maximum tank size for LNG as ship fuel is given by the round voyage distance of container ships for Europe/China trade which is 18.000, −m³. The LNG fuelled MS Trocadero in Fig. 6.2 is an example of this class of ships. There is no need for any LNG fuel tank above this size.

To enable the use of quantitative risk based distances, the IGF-Code gives an alternative to the fixed values used by Sect. 5.3.3.1 by the regulations given in Sect. 5.3.4. This alternative in Sec. 5.3.4 use factors based on damage calculation methods which include the likelihood of a damage at a defined position to the affected hull part in case of a collision. This approach originates from probabilistic damage calculation methods used by SOLAS [7].

The factor f_{CN} in Sect. 5.3.4 is the multiplication of three subfactors (comp. Fig. 8.3):

- f_l relates to the distance from aft and the tank length in order to consider collisions to the aft of the ship.
- f_t relates to the distance from side shell and the aft of the ship in order to consider side, aft collisions.

- f_v relates to the distance from tank bottom to waterline in order to consider groundings and also side collisions.

The maximum value related to cargo ships is $f_{CN} = 0,04$ and to passenger ships $f_{CN} = 0,02$.

The following explanation (Sect. 8.2.1) for the generic example may be interesting for specialists in this subject. All others might be more interested in the summarized conclusions given in Sect. 8.2.2, p. 79.

8.2.1   Generic Example for Side Collision Distances

Figure 8.4 illustrates the differences between the calculation methods. For the generic example used here the distance from the bottom is kept at 2 m and from the aft to 1/3 of the ship length. For the generic example given the f_{CN} values are calculated for the prescriptive requirement according to Sect. 5.3.3.1 IGF-Code and the probabilistic requirement according to Sect. 5.3.4 IGF-Code. Note that the three factors of f_{CN} are interlinked. For the given generic illustrative example only the distance from the side shell is varied. The other geometrical factors are kept constant or in a fixed relationship.[10]

The black line (1) represents the result of using the fixed $B/5$ value for the tank distance from the side shell up to the max value 11,5 m as given in Sect. 5.3.3.1. The distance between ship bottom shell and tank bottom is between 0,80 m up to a maximum value of 2 m for all calculations according Sects. 5.3.3.4 and 5.3.4.5 IGF-Code. The limit for passenger ships of $f_{CN} = 0,02$ is reached at a breadth of approx. 6,4 m which corresponds to a ship length of approx. 213 m (comp. graph (1) Fig. 8.5, p. 78). In the generic example case given here, longer passenger ships with higher fuel volume may fulfil the distance requirement according 5.3.3.1 of IGF-Code but will not meet the calculation result for the probabilistic approach from 5.3.4.[11] The maximum distance of 11,5 m required by 5.3.3.1 corresponds with the f_{CN} limit for cargo ships and correspond to a length of approx. 400 m (comp. graph (1) Fig. 8.5).

For the dotted and dashed blue lines (2), (3) the fuel tank distance from the side shell is set to a value which keep the limits in f_{CN}. For cargo ships the maximum value is the same as for the 5.3.3.1 calculation (11,5 m). For passenger ships (line (2)) the needed distance increase above this value for ship with a length of approx. above 310, −m (comp. graph (2) Fig. 8.5).

The comparison between graph (1) and (2) indicate that the distances using the probabilistic approach are below the calculated distances according 5.3.3.1 for small ships up to approx. 6,4 m breadth. For cargo ships (graph (3)) the values calculated according 5.3.4 are smaller if the ship length is below approx. 400 m. In other words, tanks for cargo ships using

---

[10] Breadth and length, tank volume and tank size.

[11] Please observe the note given in the coloured box below (comp. p. 78).

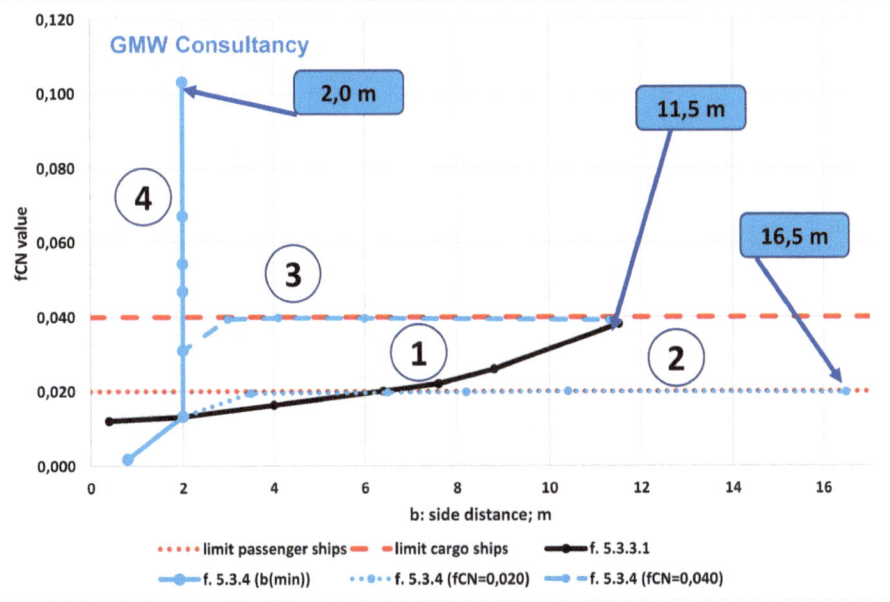

Fig. 8.4 f_{CN} values as a function of distance form the side shell according IGF-C [6]. (*Source* GMW Consultancy)

Sect. 5.4.3 requirements can be placed closer to the side shell compared to when Sect. 5.3.3.1 requirements are met.

Finally, graph (4) Fig. 8.4 illustrate the requirement for cargo ships according Sect. 5.3.3.4.2 IGF-Code which is equal to the requirements for gas carriers according IGC-Code Sect. 2.4.1.1. which limits the side distance to a maximum value of 2 m.

Figure 8.5 is intended as a summary of the rather complicated interpretation of Fig. 8.4. For the given generic example Fig. 8.5 illustrates the relations of the different f_{CN} limitations. The dotted red line indicates the max required side distance of $11, 5$ m required by Sect. 5.3.3.1 IGF-Code. The black line (1) gives the relation between the distance from side shell and ship length for the $B/5$ requirement according Sect. 5.3.3.1. The dashed dark blue line (3) shows the relationship between ship length and side distance up to an f_{CN} value of $0, 04$ for cargo ships when Sect. 5.3.4 IGF-Code is used. Above approx. $150, -$ m ship length the side distance for graph (3) is above the minimum side distance of 2 m. The side distance of 2 m is indicated by the light blue graph (4). For the $f_{CN} = 0, 02$ value for passenger ships (light blue dotted graph (2)) this minimum side distance of 2 m is reached for a ship length of approx. 70 m if f_{CN} is calculated using Sect. 5.3.4 IGF-Code.[12]

---

[12] Note again that this is an generic example with a number of conditions and NO design calculation for an individual case.

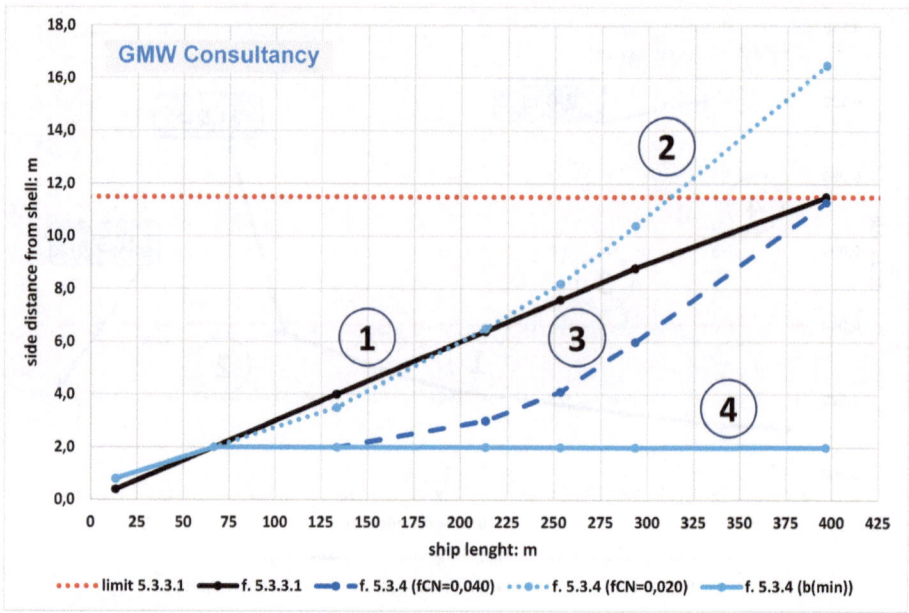

Fig. 8.5 Relation between ship length and required side distances for f_{CN} used in IGF-Code [6]. (*Source* GMW Consultancy)

With the figures from Figs. 8.4 and 8.5, it becomes obvious that the minimum distance of 2 m is exceeded for the probabilistic limits of f_{CN} according to Sec. 5.3.4 for passenger ships well below 100, −m length and for cargo ships below 150, −m length. For the limit of $f_{CN} = 0, 04$ for cargo ships, the value of 11, 5 m is exceeded for ships with a length above 400, − m.

For the explanation of the different distances comp. Figure 8.2, p. 74.

The above example (Sect. 8.2.1, p. 76) considers that the value for the factor f_v is related to the difference between distance from the fuel tank to the bottom shell plating of the ship (H) and the draft of the ship (d) (IGF-Code Sect. 5.3.4.2 $f_v = 1, 0 − 0, 8 \cdot (−(H − d)/7, 8)$). For the placement of the tank bottom below the waterline the figures become bigger than 1, 0 because the value of $−(H − d)$ becomes positive.

Related to the definition of $f_{CN} = f_l \cdot f_t \cdot f_v$ this behaviour correctly reflects the higher risk of a tank below deck to be effected by a side collision or grounding event. This risk increases with the distance below the waterline. Consequently, f_v must increase. The wording in Sec. 5.3.4.2 on limitation of the f_v factor is not in line with this probabilistic approach. It states: "...if ($H − d$) is less than or equal to 7, 8 m. f_v shall not be taken greater than 1."

To the knowledge of the author currently no official interpretation exists on how this statement has to be used. Does it mean that negative values of $(H - d)$ are not permitted? This would mean that Sec. 5.3.4 is not applicable for tanks which are partly located below the waterline and therefore not applicable for the majority of designs. Or does it mean that the maximum value used for f_v is limited to 1, 0? This would underestimate the increased risk of tank locations below the waterline.

In fact the calculation of the generic example given in Sect. 8.2.1, p. 76 with a limiting value of $f_v = 1, 0$ would give lower distances for passenger ships above approx. 150, −m and for cargo ships above approx. 400, −m compared to the values of graph (2), (3) in Fig. 8.5. This behaviour is regarded not to be in line with the probabilistic approach. For this reason, the generic example calculation has been done with the calculated values for f_v which are bigger than 1, 0 because the tank bottom is assumed to be below the waterline.

8.2.2 Summary Conclusions for Side Collision Protection

The overall conclusion from the generic example given above is that the prescriptive approach according IGF-Code Sect. 5.3.3 may simpler to use but that the probabilistic approach according to Sect. 5.3.4 of IGF-Code is better reflecting the influencing parameters for the likelihood that an LNG fuel tank is effected by a collision. It is therefore better related to the goal to design a system with equivalent safety compared to conventional designs and gives the designer more options in placing the fuel tanks.

At the same time, the 5.3.3.1 requirements do not lead in all cases to more conservative location distances compared to the application of 5.3.4. At least if the application of Sect. 5.3.4 consequently follows the idea of a probabilistic calculation approach. For this comp. the comment given in the coloured box above (p. 78).

References

1. Allianz Global-Corporate and Specialty, Safety and Shipping; Review 2023 An annual review of trends and developments in shipping losses and safety, www.agcs.allianz.com
2. internet: EXPLAINING SHIPPING-SHIPPING FACTS-Shipping and World Trade Global Supply and Demand for Seafarers; (19.07.2021): https://www.ics-shipping.org/shipping-fact/shipping-and-world-trade-global-supply-and-demand-for-seafarers/, called: 19.07.2023
3. Tanya Blake, Ines Nastali, Samira Nadkarni; IHS Markit Maritime; published by Safety at Sea magazine; The State of Maritime Safety 2020
4. https://www.spglobal.com/marketintelligence/en/; called: 19.07.2023
5. IMO (2016), IMO Resolution MSC.5(48), MSC.370(93), IMO IGC-Code, International Code for the Construction and Equipment of Ships Carrying Liquefied Gases in Bulk, IMO London, ISBN 978-92-801-1631-1

6. IMO (2016), IMO Resolution MSC.391(95), IGF-Code: International Code of Safety for Ships using Gases or other Low-Flashpoint Fuels, IMO, London, ISBN 978-92-801-1653-3
7. IMO (1974); International Convention for the Safety of Life at Seas (SOLAS), amended by resolution MSC392(95), IMO, London

Engine Room Concepts

For gas carriers and LNG fuelled ships, LNG is used as fuel in the main engines. The Codes [1, 2] both take the two concepts "ESD protected" and "inherent safe" engine room, as defined in the IGF-Code, into account.

Differing from other parts of the regulations, the two Codes do not have equivalent wording and structure. In fact the IGC-Code has general requirements in different parts of the Code for the handling of gas in rooms with possible ignition sources like compressor rooms, rooms for electrical motors, rooms for boilers, piston engines and gas turbines. No dedicated definition for inherent safe and ESD protected engine rooms is given. The interpretation of the related requirements of the IGC-Code is outlined in e.g. "Recommendations for Emergency Shutdown and Related Safety Systems" [3].

The IGF-Code [2] defines "gas safe machinery spaces" (Sect. 5.5 [2]), "ESD-protected machinery spaces" (Sect. 5.6 [2]) and details the requirements in other parts of the Code.

The requirements related to the engine room concepts have very relevant implications for the use of gas fuelled engines on board ships running with alternative fuels. Currently a controversial discussion is taking place about whether the Emergency Shut Down (ESD) concept should continue to be considered. This needs some background explanation. Therefore the engine room concepts of the IGF-Code, the background of the regulations and their implications are discussed below.

9.1 ESD Protected Engine Room

If it is not possible or adequate to protect a space where the occurrence of ignitable atmosphere can not be excluded at any time from any possible ignition source, the ESD concept is used. The ESD concept is not a special concept for ships and not only relates to engine

© The Author(s), under exclusive license to Springer Nature Switzerland AG 2025   81
G. Würsig, *The Safety Principles for the Use of Low Flashpoint Fuels in Shipping*,
Synthesis Lectures on Ocean Systems Engineering,
https://doi.org/10.1007/978-3-031-64174-9_9

rooms. It is a general safety principle, widely applied in oil, gas, refinery, chemical industry and also in shipping. For the definition of Ex-Zones, please comp. Sect. 14.2, p. 118.

When the first LNG fuelled ships went into service, they used ordinary spark ignited gas engines with fuel/air mixture before the turbocharger, as it is common for gas engines in power plants. A very good example is the MS GLUTRA [4] (comp. Figs. 1 and 1.1). To design the engine room as a Zone-1 area was not possible because the spark ignition, the generator, and the fuel/air mixing could not comply. This may have been the reason to apply the ESD principles and to name it "ESD-protected machinery space". Overall, the concept is a safe and widely used way to use engines in an environment which, in practice comply with Ex-Zone 2 requirements.

The principle of ESD engine room concept is that arrangements "...in machinery spaces are such that the spaces are considered non-hazardous under normal conditions, but under certain abnormal conditions may have the potential to become hazardous." . "...In an ESD protected machinery space a single failure may result in a gas release into the space." (Sect. 5.4.1.2 IGF-Code [2].). The two barrier principle (comp. Chap. 5, p. 29 ff.) is integrated into the shut down concept for Ex-Zone 2 equipment in the room. For Ex-Zone 2 definition comp. p. 118 in Sect. 14.2.

The concept covers normal operation conditions and failures including those with fuel release which can be expected within a reasonable likelihood according IEC-60079 [5]. Catastrophic failures according IEC-60079 Sect. 4.5 are not covered. Namely the instantaneous full bore rupture of the fuel supply line without any alarm before rupture is regarded as the major catastrophic failure which is of concern. Exactly for this worst case the ESD engine room concept requires "...explosion protection relief devices..." to prevent the collapse of the room.[1]

In practice an ESD-ER within a larger machinery space will have a "weak point" like a metal plate secured with screws which definitely will fail first and will release any overpressure into the larger machinery space. The principle is very well known from explosion flaps of engine casings. It is safe for all deflagration events. If for Hydrogen a detonation event cannot be excluded it is not safe. For a more detailed explanation of deflagration, detonation and the possibility of detonations when Hydrogen is involved comp. Sect. 4.3, p. 26.

In case of other events, all equipment not able to run under Ex-Zone 1 conditions will be shut down and de-energized to prevent any ignition. Even if the IGF-Code stipulates that an ESD-ER is "non-hazardous" under normal operating conditions, the authors recommendation is that all equipment including the electrical generators should fulfil Ex-Zone 2 requirements. Using this principle, only leaves large leaks not covered by the concept. IGF-Code name "...gas pipe rupture or blow out of gaskets...".[2] With Ex-Zone 2 equipment an ignition is unlikely even if a full bore rupture occur.

---

[1] IMO [2] 5.4.1.2 "...Failures leading to dangerous gas concentrations, e.g. gas pipe ruptures or blow out of gaskets are covered by explosion pressure relief devices and ESD arrangements."
[2] IGF-Code Sect. 5.4.1.2.

A weakness of the ESD-ER is that a shutdown caused by a gas alarm is more likely compared to an Gas-Save-ER. For this reason, redundancy is required with regard to power supply by requiring "...two or more machinery spaces...".[3] For inherent safe engine rooms the shutdown of the gas supply and a switch to oil fuel is a sufficient measure. For ESD-ER protected engine rooms a second power supply in a separate room is required. This requirement is useful because in case of a gas release into an ESD-ER a shutdown of all possible ignition sources including the engine is needed. Therefore, it is not possible to run the engine on oil fuel in case of a gas alarm.

Over time piston engine manufacturers developed concepts to protect the gas supply from ignition sources related to the engine by applying the two-barrier principle to the fuel supply. The Gas-Save-ER is the result (comp. Sect. 9.2). However, the ESD-Concept remains in practice the only concept for the use of gas turbines and fuel cell systems.

The author does not know any technical reason why this concept should not be applied. To remove it from the IGF-Code, as it is intended by some parties, would definitely be a step backwards in rule development and would open room for interpretations which most likely be more risky than the application of the dedicated ESD-Concept.

9.2   Inherent Safe Engine Room

The general concept used in industry to operate systems which cannot be designed without any ignition source is the ESD-Concept. The really particular idea in the IGF-Code therefore is not the ESD-Concept but the gas safe or inherent safe engine room concept ([2] Sect. 5.4.1.1).

The idea behind the Gas-Save-ER is that no fuel gas can enter the engine room because it is protected by a physical two boundary arrangement (comp. Sect. 5, p. 29). The protection level is, as always in the IGF-Code, IGC-Code, the protection against a single failure. In this case that "...a single failure can not lead to a gas release into the machinery space" (comp. Sect. 5.5.1 and for technical details Sect. 9.6.1 [2]). Note that the two barrier principle is required "...until gas is injected into the" injection "chamber".[4]

Figure 5.1 illustrates the principles given in Sect. 9.6 [1] for gas-safe machinery spaces. The regulations related to double walled piping and ventilated ducts in the IGF-Code are discussed in Sect. 5.1, p. 29.

---

[3] Section 5.6.3.1, IGF-Code [2].
[4] Section 9.6.2 [2].

References

1. IMO (2016), IMO Resolution MSC.5(48), MSC.370(93), IMO IGC-Code, International Code for the Construction and Equipment of Ships Carrying Liquefied Gases in Bulk, IMO London, ISBN 978-92-801-1631-1
2. IMO (2016), IMO Resolution MSC.391(95), IGF-Code: International Code of Safety for Ships using Gases or other Low-Flashpoint Fuels, IMO, London, ISBN 978-92-801-1653-3
3. SIGTTO Recommendations for Emergency Shutdown and Related Safety Systems, Second Edition 2021, ISBN: 978-1-85609-998-1, eBook ISBN: 978-1-85609-999-8, Witherby Publishing, Livingston, Scotland, UK, 2021
4. Einang, Per Magne; Haavik, Konrad Magnus; The Norwegian LNG Ferry; PAPER A-095 NGV 2000 YOKOHAMA
5. Explosive atmospheres-Part 10-1: Classification of areas-Explosive gas atmospheres; IEC Standard 60079-10-1, 2021,INTERNATIONAL ELECTROTECHNICAL COMMISSION (IEC), ISBN 978-2-8322-9213-6

Tank Connection Space

The most relevant space with regard to safety is the tank connection space. The tank connection space includes all tank connections of the fuell tank including the first valve outside of the tank (tank connection space). The safety aspects related to this space are discussed below. For illustration an example to illustrate the possible scope of release from piping installed at the tank top is given in Sect. 10.2, p. 87. Other special spaces defined in the IGF-Code are listed at the end of this Chapter (Sect. 10.3, p. 89).

Details on the tank connection space definition are given in Sect. 6.6.1 which includes the definition in the IGF-Code (p. 51). Figure 6.14 illustrates the concept.

10.1 What is a Tank Connection Space?

A failure of the tank connection space if related to a tank bottom line will lead to the release of the complete tank content. With a small leak this might need weeks but is not or at least very difficult to mitigate.[1]

The potentially unlimited release of alternative fuel is the "unique selling point" of the tank connection space. The other spaces do not have this risk potential.

The idea of the tank connection space as defined in Sect. 2.2.15.3 of the IGF-Code [1] is closely related to the safety principles for type C gas carrier cargo tanks and the deviation from the principle that no connections should be below the liquid level. For details comp Sect. 6.6.1, p. 50.

Currently the most prevalent understanding today is that the tank connection space is the room where the first tank valves are installed as can be concluded e.g. from Sect. 13.4 of the

---

[1] For discussion on the criticality of bottom lines comp. Sect. 6.6.1.

© The Author(s), under exclusive license to Springer Nature Switzerland AG 2025   85
G. Würsig, *The Safety Principles for the Use of Low Flashpoint Fuels in Shipping*,
Synthesis Lectures on Ocean Systems Engineering,
https://doi.org/10.1007/978-3-031-64174-9_10

IGF-Code.[2] This is correct with respect to the possibility of releasing the tank content into this room if the first valve on a liquid line connected to the tank bottom fails.

It need to be noted that also the space between tank connections and this room is part of the tank connection space. This is the meaning of the definition of the tank connection space which include this area upstream of the first valves.[3] To say it in polite words, it must be admitted that this definition is not very clear. The IMO Guidelines [2] defined the space now called "Tank connection space" as "Tank room".[4] Also this definition is not very clear.

The original motivation for both definitions is the assumption that it is possible to have a large leak in the piping.[5] For this reason, the outer shell of the early LNG fuel tanks with vacuum insulation had a stainless steel outer shell and a thermal insulation of the hold space walls.

The failure scenarios for the piping upstream of the first valves, the first valves and downstream of these valves are different. Therefore, the author is convinced that the two spaces should be handled different and should have separate a definitions.

A possibility is to define the space between pipe connection to the tank and the inlet flange of the first valve as tank connection space and the space where the valves are installed as "tank valve compartment".[6] E.g., the requirements in Sect. 13.4 would become more clear. The proposal to differentiate between these two spaces was not successful. The result is that a lot of IGF-Code users are confused when they need to know what a tank connection space is. It could well be that most users think it is only the "tank valve compartment" as proposed above.

This remark is related to origin of the tank connection space definition which is, for the author, related to assumed leaks from bottom lines of pressure vessels of IMO International Maritime Dangerous Goods Code (IMDG-Code) LNG containers even if the pressure vessels themself could have been qualified as type C tanks.

The two barrier type C tank concept (comp. Sect. 6.5, p. 46) is not known for shore installations. Piping connected e.g. to a vacuum insulated container with carbon steel outer shell for UN Type 2.1 cargo according the IMDG-Code[7] have bottom connections. The pipes and the containers do not fulfil the type C design requirements. The pressure vessels of these containers can be qualified as type C tanks to use them in LNG fuelled ships as fuel tanks but the piping can not. The view of the author is that

---

[2] "13.4 Regulation for tank connection space", "13.4.1 The tank connection space shall be provided with effective mechanical forced ventilation...".

[3] 2.2.15.3 IGF-Code [1]: "Tank connection space is a space surrounding all tank connections and tank valves that is required for tanks with such connections in enclosed spaces."

[4] IMO Guideline 2009 [2]: "1.3.33 Tank room means the gastight space surrounding the bunker tank, containing all tank connections and all tank valves."

[5] Most time a full bore rupture is stipulated.

[6] Or as "tank valve space" if a tank dome on deck is the location for the installation of the valves.

one reason for the introduction of the tank connection space concept was to enable container solutions to be used for LNG as ship fuel tanks.

It should be noted that the IMDG-Code do not permit the connection of the transported containers to the ship system! This is one difference between an IMDG-Code Container for liquefied gases and a liquefied gas tank e.g. used in process systems on board of a ships as storage tank for liquefied gases.

It should also be noted that exactly these type of IMDG-Code containers are today proposed for Carbon Capture and Use (CCU) systems on board for CO_2 storage. The same has been done for LNG storage as fuel in the early days of LNG as ship fuel. The wrong argument is the same in both cases: "The container is the same as a type C tank and needed to overcome the missing bunkering infrastructure!". By the way to counter measure this argumentation was the reason to implement Sect. "6.5 Regulations for portable liquefied gas fuel tanks" into the IGF-Code (comp. Sect. 6.6.2, p. 53).

10.2   How Much Gas Can be Released from a Tank Connection at the Tank Top?

If the leak from a tank is not related to a tank connection below the liquid level but to a leak in a line from the gas space of the tank, the flow can be stopped by reducing the pressure difference between gas pressure in the tank to the ambient pressure. The following generic example may illustrate the amount of potential release.

Assuming a type C tank with a total geometrical volume of 1.000, $-m^3$ and filling level of approx. 50%. This give a tank content of approx. 200 t of LNG. Assuming a uniform temperature in the complete tank and a pressure of approx. 4 bar g allow to calculate the necessary amount of evaporation to reach equivalence with ambient pressure by a simple stationary heat balance.

The result is that approx. 26 t of fuel gas must be relieved to reach pressure equilibrium between the tank and ambient.

To calculate the time needed to reach pressure equilibrium with ambient it is assumed that the release is equivalent to the capacity of the two Pressure Relief Valves (PRVs) from Fig. 7.1, p. 65 (Sect. 7.2.2)[8] and that the flow is constant for the complete time[9]. In this case the time until pressure equilibrium is reached will be approx. 1 h and 20 min.

Of course, smaller tanks will give lower figures. For bigger tanks a lower pressure has to be assumed because they will be type A, B or membrane tanks. This might give lower

---

[7] UN Type 2.1 products are liquefied gases.

[8] Approx. two $DN - 150$ pipes.

[9] Mass flow is reduces when the tank pressure drops below the De Laval pressure which is approx. 2, 5 bar tank pressure in the given case

figures, but it must be considered that the volumes of these tanks will be much higher. At the end the named figures may give a feeling for the scope of the event.

As all example figures given in this publication, this example indicate the scope of the event. It do not give a practical release figure.

What else can be learned from the above generic example:

First of all the amount gas released also indicates what happens if a PRV is fails and the flow can not be stopped (comp. Sect. 7.2.2, p. 65). The relief area may "only" be the area of one PRV. This would not affect the amount of relieved gas. It will only double the relief time.

Moving from "unhappy" incidents towards something positive leads to the possibility to illustrate how liquefied gas storage works. If the tank is filled to 50% with LNG at boiling conditions to $p_{s0} = 1$ bar (Note: bar is always absolute pressure!), $T_{s0} = 111$ K and the maximum equilibrium is as given for the example at $p_{s1} = 5$ bar, $T_{s1} = 135$ K the energy difference stored in the liquid and gas is related to the increase of mass and temperature in the gas space and the temperature increase in the liquid LNG.

Assuming a BOR of 0, 3%/d gives a holding time of approx. 29 d until the 5 bar pressure is reached. Approx. 91% of the related energy increase in the tank is caused by the temperature increase in the liquid, 8% by the evaporation of liquid to meet the mass balance in the gas space and only 1% is due to the temperature increase of the gas.

Note that this range of holding time only works with a forced cooling by spraying cold liquid into the gas space and some mixing of the liquid phase. The additional pump power will reduce the holding time. Without this temperature maintenance the holding time will be reduced significantly because the liquid will not absorb much heat, the evaporation is very limited and the gas in the upper part is superheated but can not absorb much energy.

At the end this example also has an unhappy message for those who like to store cooled gas instead of liquefied gas because the liquid phase stores more than 90% of the energy. The holding time would decrease by at least the factor of 10 if no liquid but cold gas is stored!

10.3 Other Special Spaces

Other special spaces on board are:

- The fuel preparation room (comp IGF-Code definition Sect. 2.2.17) is an enclosed space containing process equipment for fuel preparation. If not located on open deck, the fuel preparation room needs to meet the requirements for tank connection spaces (Sect. 5.8 IGF-Code).
  Note that such equivalences lead to more confusion about the question "what is a tank connection space" because the most relevant feature of the tank connection space is that it has an unlimited release source (the fuel tank) as the risk. Compared to this, the possible fuel spill into a fuel preparation room is nearly "nothing"!
- The Gas Valve Unit (GVU) is not defined in the IGF-Code but is nevertheless needed on board of all ships with alternative fuels.[10] It is located close before the engine and contain the process equipment for fuel pressure regulation of the engine. A short distance to the engine ensures that the amount of gas upstream of the last process valve before the engine is small. With this the reaction time of the engine for load changes is kept at a low value.
  The gas valve unit can be designed as a part of the fuel supply line in the engine room or as a separate room designed equivalent to the fuel preparation room.
- Semi-enclose spaces are spaces on deck which are partly covered by the ship structure. The bunker station shown in Fig. 10.1 installed on a cargo vessel is an example.
- Bunker stations in enclosed spaces. On ferries and cruise ships the bunker stations are installed in the ship. Note that these stations do not create any additional risk.[11] The opposite is true. A well designed bunker station inside the ship which has an adequate suction ventilation avoids gas clouds outside of the station if a spill inside of the station occurs. In this regard, it mitigates the risk for the surrounding of the bunker station.
  Note that these kinds of bunker stations should not be manned when fuel transfer takes place. The reason is that persons in the room can not escape in case of a large spray of cryogenic liquid like e.g. LNG. If NH_3 might become a ship fuel the situation would be worse.[12]

---

[10] Note that different names/shortcuts for this process unit are common (e.g. GVT, Gas Regulation Train,...).

[11] If designed with know how.

[12] The author has seen a number of bunkering operations where the operators stayed in such a bunker station or at least were present until the full flow was established. By the way, the first flow is a critical moment for leaks to occur.

Fig. 10.1 Semi enclosed bunker station on a cargo ship. (*Source* Dr. Gerd Wuersig)

References

1. IMO (2016), IMO Resolution MSC.391(95), IGF-Code: International Code of Safety for Ships using Gases or other Low-Flashpoint Fuels, IMO, London, ISBN 978-92-801-1653-3
2. IMO RESOLUTION MSC.285(86), adopted on 1 June 2009, INTERIM GUIDELINES ON SAFETY FOR NATURAL GAS-FUELLED ENGINE INSTALLATIONS IN SHIPS, London, 2009

Fuel Bunkering

11

Apart from the fuel tank itself, the fuel bunkering is the potential source of the largest amount of fuel release.[1] At the same time, the frequency of a possible failure occurrence is high because bunkering is done frequently and the number of total bunkering operations is increasing with the number of ships with alternative fuels (comp. Chap. 2, p. 9). The risk related to refuelling ships can be compared to refuelling aircraft. The fuel amount may be larger but on the other hand the distances between the planes on an airport is much smaller compared to the distances of the ships in a harbour.[2]

As explained earlier (comp. Sect. 6.1, p. 35), the fuel tanks for LNG as fuel are much smaller than the LNG tanks of LNG gas carriers but in most cases significantly larger than road or rail tanks for LNG transport. The range can be assumed to be between 400, $-m^3$ to 20.000, $-m^3$ (comp. p. 35).

The largest tanks which have been built for LNG as fuel have a capacity of approx. 18.000, $-m^3$ (comp. Fig. 11.1). Tank sizes above 20.000, $-m^3$ are not needed.[3] Today's standard LNG-carriers with approx. 170.000, $-m^3$ in most cases have a single tank LNG capacity well above 40.000, $-m^3$.

---

[1] IGF-Code Sect. 8 gives the relevant regulations for LNG as fuel; note that 8.1–8.3 can be regarded to be applicable for all alternative fuels (comp. Sect. 1.2, p. 3 for the reasoning).

[2] It need to be considered that a plane has no relative movements between plane and bunker station during bunkering. A ship has such relative movements.

[3] Comp. Sect. 6.1.

© The Author(s), under exclusive license to Springer Nature Switzerland AG 2025
G. Würsig, *The Safety Principles for the Use of Low Flashpoint Fuels in Shipping*,
Synthesis Lectures on Ocean Systems Engineering,
https://doi.org/10.1007/978-3-031-64174-9_11

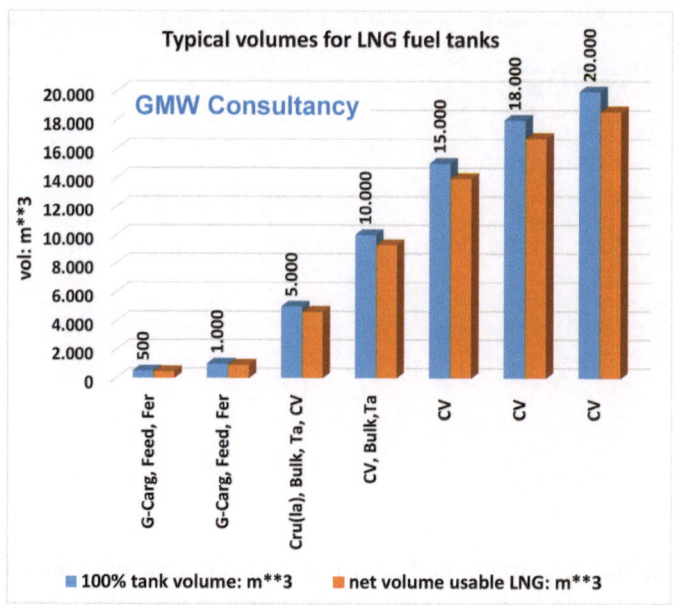

Fig. 11.1 The figure illustrates that even the largest LNG fuel tanks are less than half the size as common LNG carrier cargo tanks. (*Source* GMW Consultancy) Ship types: G-Carg = General Cargo; Feed = Feeder CV; Fer = Ferry; Cru(la) = Crude oil tanker; Bulk = Bulk Carrier; Ta = Tanker; CV = Container Vessel

Typically LNG cargo transfer to/from LNG carriers takes place with a rate of up to approx. 4.000, $-\mathrm{m}^3/\mathrm{h}$. For LNG fuelled ships the rates are up to approx. 1.000, $-\mathrm{m}^3/\mathrm{h}$ as a maximum. Therefore the transfer hard-arms or hoses and related connections for LNG as fuel can be significantly smaller compared to the hard-arms or hoses for LNG as cargo which have a diameter of $16^{in} \approx DN\ 400$ in most cases.

This Section highlights some of the most relevant aspects for bunkering LNG as fuel.

Note that for future fuels like Methanol, LH2 or possibly Ammonia the hoses/hard-arms and couplings need to be increased if the same transfer time as that for LNG is stipulated. Assuming the same energy requirement, the factor for volume flow increase or alternatively fuelling time increase will be 2, 4 for LH2, 1, 6 for Ammonia and 1, 4 for Methanol.

PtX-FT will have only a volume flow or time for fuelling equivalent to 0, 6 of the related value for LNG or Power to X Liquefied Methane Gas (PtX-LMG) as fuel.

11.1 Coupling

IGF-Code Sect. 8.4.1 requires the use of dry-disconnect couplings in combination with dry break-away couplings. The principle of the dry-disconnect coupling required by Sect. 8.4.1 of the IGF-Code is illustrated by Fig. 11.2 and the required dry break-away coupling by Fig. 11.3. Couplings working according to the same principle are widely used, e.g. for refuelling of aircraft.

The current IGF-Code (2016 edition) does not allow other types of couplings. This was discussed intensively during the development of the Code. The basic reason for limiting the types of couplings is that the high number of refuellings require an as simple as possible procedure with limited possibilities for things to go wrong. This is the only way to limit the frequency and effect of malfunction during connection/disconnecting and operation of bunker connections. More details about the reasoning to exclude other types of couplings are given in the following.

> Note that Sect. 8.4 IGF-Code has been revised. The revision is scheduled to be adopted by IMO MSC 108 which will take place in May 2024 (note that this text was finished in the beginning of 2024). The formal enforcement will be in 2026 at the latest. If a safety analysis is done according to Sect. 4.2.2 of the IGF-Code, it will be also allowed to use manual connect or hydraulic connect couplers (these are the types used for LNG as cargo transfer with DN-400 pipe/hose diameter). It will also be permitted to use "a bolted flange assembly", as it is common for conventional oil fuel today.

Dry Quick Release/Disconnect Couplings

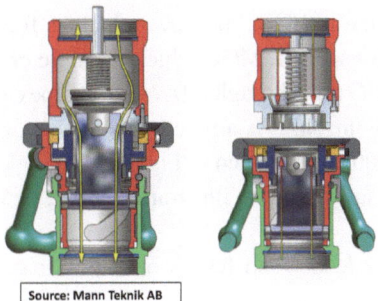

- Flow path opened as part of connection process
- No acess of air into the system
- Prevent spillage.
- Keep hazardous liquids and vapors in-line
- Connect and disconnect under pressure and flow
- Remove human error elements
- Currently available up to 8 inch

Source: Mann Teknik AB

Fig. 11.2 Dry Quick Release/Disconnect Coupling (DQRDC). (*Source* MannTek)

**Excluding hose ruptures by movements
- Dry Break Coupling -**

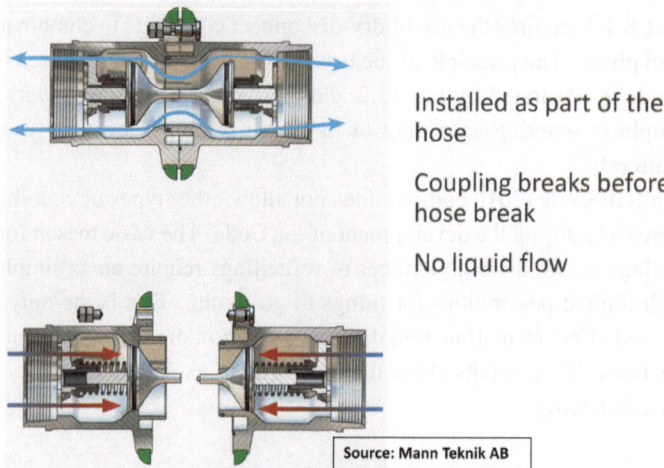

Installed as part of the
hose

Coupling breaks before
hose break

No liquid flow

Source: Mann Teknik AB

Fig. 11.3 Dry break coupling illustrating the principle how the release is prevented in case of disconnection with bunkering connection still in place. (*Source* MannTek)

Figure 11.3 illustrate the working principle. A version needing forces on the connection to break the screws of the flange is shown. IGF-Code conform versions have an automated release design which can be operated remotely and is available from the same manufacturer.

The Dry Quick Release/Disconnect Coupling (DQRDC) as illustrated in Fig. 11.2 can be operated by hand up to a size of $6^{in} \approx DN-150$ relatively easily. The operation by hand is limited by the weight of the coupling. They are available up to $8^{in} \approx DN - 200$. Operation at this size may need some support structure. As illustrated below a 6^{in} coupling is completely sufficient, at least for LNG.

Figure 11.4 gives the relation between the pipe diameter, the volume flow through the pipe and the related flow velocity.

Related to the physical properties of the medium, the flow velocities for LNG can be significantly higher than that for conventional HFO, MGO which are more or less the same as for water. According to Fig. 11.4 the LNG flow through a $DN - 150$ pipe can be between approx. 380, $-m^3$/h and approx. 1.000, $-m^3$/h.[4] As a practical reference, the values for DQRDC of MannTek DN-150 and DN-200 are indicated in Fig. 11.4 [1].

The required flow duration for typical tank sizes are illustrated by Fig. 11.5. The related tank sizes are given by Fig. 11.1.

Typical general cargo ships, container feeder and ferries operating, e.g. in European waters, which have only short port stays may have a tank capacity between 500, $-m^3$ to

---

[4] Doted green, yellow line in Fig. 11.4.

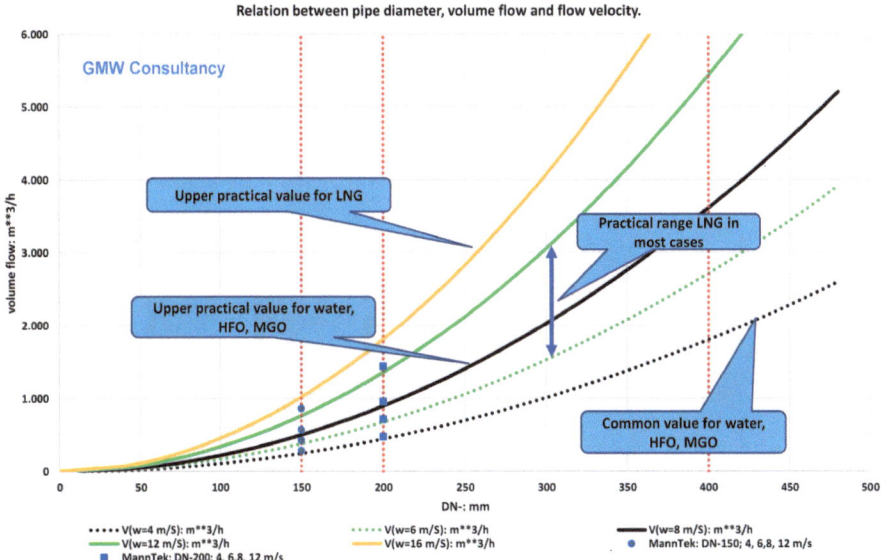

Fig. 11.4 Illustration of volume flow related to pipe diameter and flow velocity. (*Source* GMW Consultancy)

1.000, $-m^3$. Compare (1), (2) in Fig. 11.5. For these ships hose diameters of DN-80 to DN-100 are sufficient to meet the refuelling times assumed for Fig. 11.5.[5]

Large Cruise ships and small Bulkers, Tankers and Container ships in international trade may have a tank capacity of approx. 5.000, $-m^3$ LNG (comp. (3)). Port stay may be a limiting factor for large Cruise ships. Practically the port stays are normally between 8 and 10 h.[6] For ships of up to this size, the dominant tank type can be assumed to be the type C pressure tank.

The lower end of the large Container ships (8.000, $-$TEU to 10.000, $-TEU$), large Bulker and Tankers may have tank volumes of approx. 10.000, $-m^3$ (comp. (4)). Usually these ships stay in port for more than one day in most cases.[7]

Larger tank sizes are related to large container ships with 15.000, $-TEU$ (comp. (5)) and above (comp (6), (7)) in Europe/Asia, Europe/Americas or Americas/Asia trade. The largest container ships currently running on LNG have 23.000, $-TEU$ (comp. (6)).[8] These ships stay several days in harbour in most cases. Refuelling is needed once per round voyage,

---

[5] (1): DN-80, $w = 6, 4$ m/s; (2): DN-100, $w = 8, 2$ m/s.

[6] (3). DN-150, $w = 12, 2$ m/s.

[7] (4). DN-150, $w = 14, 6$ m/s.

[8] (5), (6), (7): DN-150, $w = 15$ m/s.

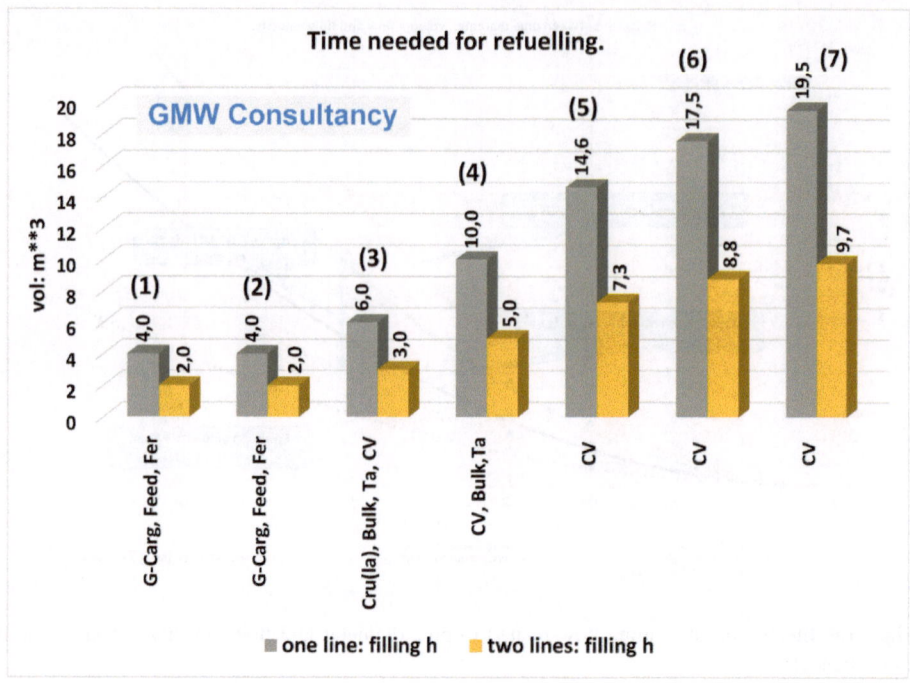

Fig. 11.5 Illustration of the required filling time for typical tank sizes. (*Source* GMW Consultancy)

which today typically is about 90 days to 100 days. Practical observed[9] refuelling times are 20–30 h with related port stays of approx. 40–60 h.

The figures above relate to average volume flows and the refuelling time does not consider the bunker ship mooring times and the time for connection setting. The latter is significantly reduced when compared to a flange connection. For DQRDC this time can be calculated in minutes and must be calculated in hours for flanged connections. As the DQRDC is a closed system the drying and inerting procedures technically necessary for flange connections are technically not needed for DQRDC.[10]

Please note that the above calculations are simplified calculations which specially do not consider the flow limitations which might be caused by flashing of LNG. If the tank is not properly prepared for bunkering by spray cooling this effect can be most relevant for prismatic tanks with a pressure limitation well below 2 bar.[11] In practice type C tanks have an advantage in this regard because they can handle the flash gas by pressure increase. The practical filling method for type C tanks is top filling which ensures the cooling down of the gas space and related tank structure.

---

[9] By the author.

[10] Note that regulations require inerting.

[11] Absolute pressure.

The author's conclusion from the above discussion is that in practice hose sizes and related tank connections beyond DN-150 or in seldom cases DN-200 are not needed for LNG refuelling. Given the author's background knowledge which include the results of evaluations from known incidents, the author is still sure that the employment of DQRDC is the best way to ensure safe bunker connection, disconnection and operation for ships with alternative fuel.

Reference

1. Flow calculation with friction loss (breakaway version), DN-150, DN-200 MannTek, 2019

Tank Filling Levels

12

The maximum tank filling is since decades a matter of discussion and battle between safety matters and commercial interests. The question always is how much is too much? The author experienced that even process engineers do not have a proper understanding of the process technology related to filing limits often even if they are claimed to be experts on the subject.

For a clear understanding, the basics for liquid and liquefied gas expansion is first explained. The regulations for bulk liquefied gas transport and for LNG as fuel and other alternative fuels are explained and discussed. Finally, the conclusions for a safe filling restrictions are given.

12.1   Why Not Fill a Tank Completely

Almost all liquids and also liquefied gases expand when the temperature is increased at a constant pressure.[1] E.g. if water is heated at the bottom of a pot, the warm water is rising up because the density decreases with increasing temperature and the volume of warm water rises according to the Archimedes principle. Another example is the expansion tank of a water heating which is compensation the higher volume of the hot water.

In the following the author uses water as a reference fluid because the physics for expansion[2] and evaporation are the same as for oil fuels and liquefied gases. The advantage of

---

[1] The "nearly" is for the author mainly related to the anomaly of water around the freezing point at 1 bar.

[2] At temperatures well above the freezing point at 1 bar.

© The Author(s), under exclusive license to Springer Nature Switzerland AG 2025   99
G. Würsig, *The Safety Principles for the Use of Low Flashpoint Fuels in Shipping*,
Synthesis Lectures on Ocean Systems Engineering,
https://doi.org/10.1007/978-3-031-64174-9_12

water in this context is that everyone knows it and can do his own "tests" without the risk of ignition.[3]

12.1.1 Liquefied Gas is Boiling all the Time

If not used for cooking water is most times used below the boiling temperature. It is used as sub-cooled liquid. A liquefied gas is transported and used more or less at its boiling point and therefore comparable with water at the boiling point.

If water is heated in an open pot to boiling temperature at atmospheric pressure[4] the vapour can escape to the ambient and the pressure is kept constant. If a pressure cooker is used the water has a pressure above ambient temperature and the corresponding boiling temperature is higher. Figure 12.1 illustrates this and give the related gas carrier ship types.[5]

Fig. 12.1 Analogy between water heating and liquefied gas transport (*Note* Type A,B, Membrane tanks have a maximum pressure of 0, 7 bar g above atmospheric pressure). (*Source* GMW Consultancy)

---

[3] Anyhow, some care should be taken if doing tests with hot water specially if it is under some pressure!

[4] Approx. 1 bar.

[5] The some hundred *mbar* pressure increase in the Moss type LNG carrier is neglected here.

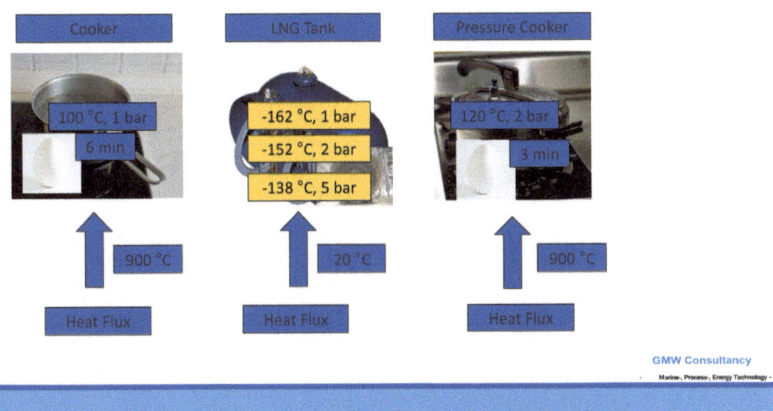

Fig. 12.2 The temperature level is the key difference between water boiling and LNG boiling: Analogy between water heating and LNG storage in Type C tanks. (*Source* GMW Consultancy, Type-C tank image: TGE-Marine)

The temperature of water in the pot stays constant until all water is evaporated, if the pressure is kept constant. As shown in Fig. 12.2, the fact that this is also the case for pressures above ambient pressure is the reason why a pressure cooker is used. Higher boiling temperature related to higher boiling pressure lead to shorter cooking time.

As illustrated in Fig. 12.2 the heat source in LNG storage is of course not the heat from the stove it is the ambient which is always much warmer than the LNG in the tank.[6]

Beside the volume expansion of the liquid with increasing temperature, the energy needed to increase the temperature and pressure must also be considered. With increasing temperature and pressure, the energy needed to increase the temperature and pressure by a given Δ goes down.

The boiling pressure and liquid temperature increase per energy unit is moderate up to some distance from the thermodynamic critical point because the energy absorption by the liquid is buffering the pressure increase. The critical point for LNG is $p_{crit} = 46, 26$ bar, $T_{crit} = 190, 8\ K = -82, 4\ ^{\circ}C$. Close to p_{crit}, T_{crit} and above the pressure increase per energy unit is increasing. The same amount of added energy lead to higher pressure increase. As a rule of thump, a maximum storage temperature in Kelvin should not be higher than 80% of the critical value. For Methane these 80% value are $T_{s,max} = 152\ K = -121\ ^{\circ}C$, $p_{s,max} = 11, 7$ bar $= 10, 7$ bar g.

---

[6] Note that the temperature assumed for the fire case to size the PRVs (comp. Sect. 7.1.2, p. 62) is approx. at the level indicated for the stove in Fig. 12.2.

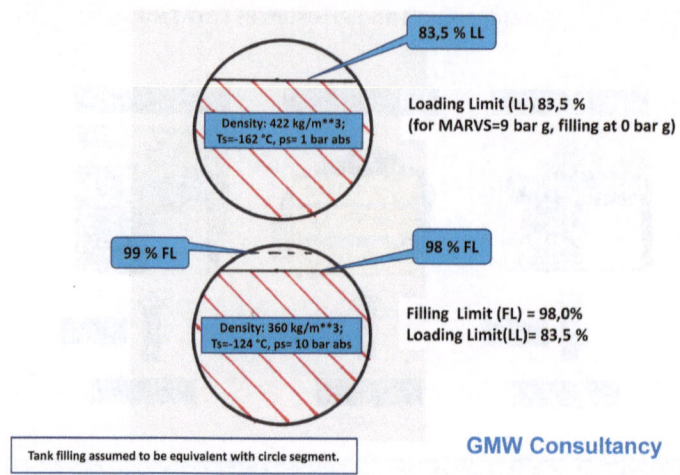

Fig. 12.3 Illustration of Loading Limit (LL) and Filling Limit (FL) for a cylindrical type C LNG tank. (*Source* GMW Consultancy)

12.1.2 Liquefied Gas Expands When Heated to Higher Boiling Pressure

The basic idea behind the regulation for tank filling limits is that the heat from the ambient is completely absorbed by the tank content and that the tank content has a uniform temperature. For liquefied gases it is assumed that temperature and pressure are boiling temperature and pressure. Figure 12.3 illustrate the principle for a cylindrical Type C LNG tank.[7]

When the MARVS pressure is reached the tank should not be liquid full. The general maximum value is in IGF-Code and IGC-Code 98% of the tank geometrical volume. The IGC-Code allows higher filling limits under consideration of special trim and list conditions.[8] Practically most LNG carriers with membrane tanks have filling limits of approx. 99%.

The illustration given in Fig. 12.3 is related to a tank with a maximum FL of 98%. This level would be reached at MARVS pressure assuming the theoretical expansion behaviour as described above. The maximum loadable cargo or ship fuel is given by the *tank Loading Limit (LL)* which is illustrated for a loading of LNG at 1 bar pressure and the related boiling temperature.

A uniform temperature of the liquid in the tank which follows the ambient temperature is nearly reached for liquids like oil, Methanol or water. For liquefied gases, the above assumptions are useful to ensure a reasonable limitation of filling but they do not reflect the

---

[7] Note: The height of the liquid level depends on the tank shape. The absolute distance to the tank top is smaller for prismatic tanks because the relation between tank filling and liquid level is linear which is not the case for a cylindrical tank.

[8] IGC-Code, Sect. 15.4.1 [1].

real behaviour of the tank content. The principle behaviour for liquefied gases is explained in the following section (Sect. 12.1.3, p. 103).

12.1.3 Tank Pressure and Boiling Pressure are Related But Not the Same

Most engineers are confident that for liquefied gas tanks on seagoing ships the relation between heat flux from ambient, pressure and temperature is exactly as described in Sect. 12.1.2. The reasoning is the assumption that the ship movements always ensure a movement of the tank content and with this a good mixing. For the benefit of safety this is not the case.

Figure 12.4 gives the pressure and temperature behaviour of an Ethylene filled type C deck tank with approx 100 m^3 volume on a gas carrier. The results are part of a measurement campaign which included 5 ships and 2 Ethylene, 2 Propylene and 1 Butane tanks. The results shown on Fig. 12.4 are typical.

The reference used for the pressure is the boiling pressure related to the liquid temperature in the tank. The reference for the temperature is the boiling temperature related to the measured tank pressure in the gas phase. The values above 1, 0 indicate that the pressure in

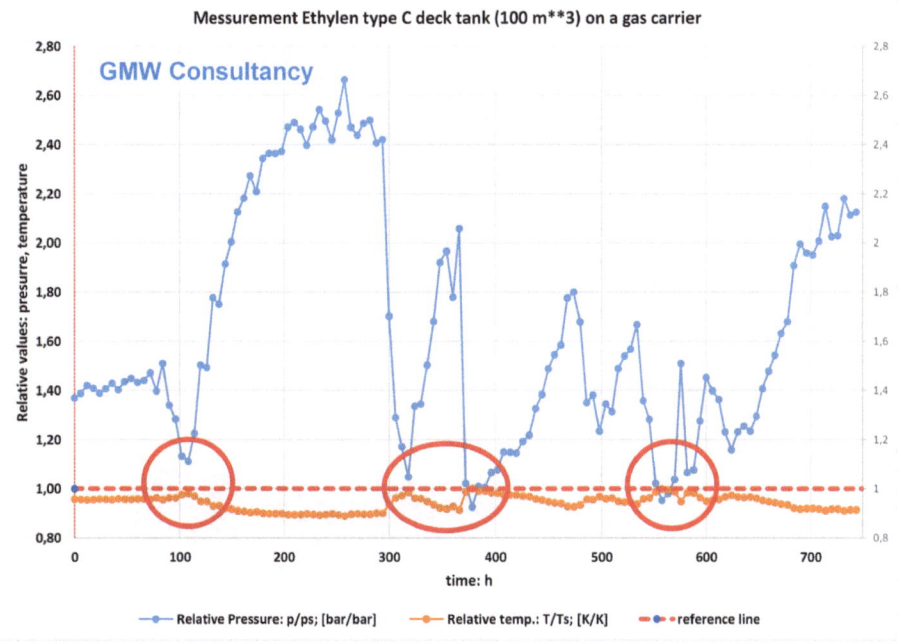

Fig. 12.4 Pressure and temperature deviations from boiling point (on board measurement for an Ethylene deck tank). (*Source* GMW Consultancy)

the tank is above the boiling pressure. For the temperatures the values below 1, 0 indicate that the temperature is below the boiling temperature related to the tank pressure. A uniform energy distribution in liquid and gas phase as explained before (comp. Sect. 12.1.2, p. 102) would lead to constant values for relative pressure and temperature of 1, 0 as indicated by the red reference line in Fig. 12.4.

The lowest point of the relative pressure in the red circled areas in Fig. 12.4 indicate the points when the re-liquefaction was switched off. The area just before is the time when the re-liquefaction was running. In between the re-liquefaction was switched off. The re-liquefaction was switched off most of the time.

The vapour phase is clearly superheated and the liquid phase clearly sub-cooled most of the time when the re-liquefaction is not running. A continuous mixing would lead to values of approx. 1, 0. This does not occur often. Nearly equilibrium or even sub-cooling is only reached just before the re-liquefaction is switched off (red circled areas Fig. 12.4). Some mixing by ship movements is indicated by the variations during the time when the re-liquefaction was switched off.

Specially the area between 100 h and 300 h indicate that the tank pressure is much higher than the related saturation pressure.[9] Consequently the liquid temperature is much lower than the related boiling temperature.[10]

The on board measurements discussed above confirm systematic measurements which have been performed with a 61 m^3 LH2 tank in 1996 in Germany by Germanischer Lloyd and BAM (comp. Fig. 12.5).

Fig. 12.5 LH2 test tank on BAM test area in Horstwalde, Germany 1996. (*Source* Dr. Gerd Wuersig; author: the grey container was the author's "home" for summer and autumn of 1996)

---

[9] $\Delta p \approx 3, 8$ bar above saturation pressure.
[10] $\Delta T \approx 22\ K$ below boiling temperature.

The results are evaluated, e.g., in the Diploma work of Peter Hauschildt [2]. The measured pressure increase in the LH2 tank was approx. 30% faster compared to the results from the assumptions used in [1, 3] and explained in Sect. 12.1.2, p. 102.

> LH2 is an ideal medium to evaluate the temperature pressure behaviour of liquefied gas storage because most additional parameters which might influence the results are not present. Firstly, the buffer capacity of the inner tank shell is negligible because the heat capacity of the steel drops to zero at the low temperature of LH2 (20 K at 1 bar). Secondly, the heat capacity of the vacuum foil insulation is also very low. Thirdly, LH2 is a very pure medium without impurities which might influence the convection in liquid and gas space. Anyway, except for Helium, all impurities would be solid at LH2 temperature.

The real behaviour described above leads to a reduced liquid expansion and a faster pressure increase compared to the simplified model used to define the maximum LL, FL (comp. Sect. 12.1.2, p. 102). In other words, the tank pressure reaches MARVS before the tank can get liquid full. For low pressure type A, B and membrane tanks the rise in liquid level is limited by the transport pressure to a maximum of 0, 7 bar g according to IGC-Code and IGF-Code. In practice the normal transport pressure is approx. 250 mbar g. Keeping this pressure by BOG removal from the tank in LNG transport, or by re-liquefaction for other liquefied gases, limits the possible liquid level increase. This is the reason why very high filling limits do not compromise safety in liquefied gas transport.

12.1.4 Preventing Overfilling and PRV Acting in Incidental Situations

The safety philosophy behind the IGC-Code, IGF-Code regulations assume the prevention of tank overfilling in two cases:

1. The cargo/fuel pressure and temperature control by pressure keeping or re-liquefaction.
2. A fire which partly engulfs the tank/ship as explained in Sect. 7.2.2, p. 65.

12.1.4.1 Liquid Expansion Under Normal Ambient Temperature Conditions

IGF-Code, IGC-Code require technical means to keep the pressure within the design limits if the tanks are not designed to allow a temperature increase up to the upper ambient temperature which is given with 45 °C air temperature and 32 °C sea water temperature. A time for which this pressure control should work is not defined in the IGC-Code. For liquefied gas carriers the requirement of 21 d set by the US in [4] is the generally accepted design criteria. For gas fuelled ships IGF-Code Part A1, Sect. 6.9.1.1 stipulates that the pressure must be kept "below the set pressure of the tank pressure relief valves for a period of 15 days".[11] From this requirement it is clear that any LNG gas carrier and any LNG fuelled ships has adequate means to kep the pressure without venting natural gas to atmosphere.

For gas fuelled ships re-liquefaction system are not common. In general the pressure is kept within the design limits by the use of the fuel in the ship machinery system. For cases when the machinery system is not operable[12] a secondary means of pressure control is required. In most cases, this is a burner enabling the combustion of boil off gases, or for type C tanks an adequate insulation to maintain the pressure for the named 15 d. For passenger and cruise ships it is also acceptable to use the needed auxiliary energy as a second mean of pressure control. The reader may note that there are no requirements for the redundancy of the consumers. It may be acceptable when, for instance, all consumers are located in one engine room and therefore can fail at the same time.

From the above (Ec. Sect. 12.1.3, p. 103) it is obvious that for type C tanks an active pressure management is needed or at least useful to control the pressure. This can be done by pumping liquid from the tank and spraying it back into the vapour space of the tank. By such means the pressure is kept at the boiling pressure of the liquid. The tank energy balance including the additional heat from the pumping system should ensure that the pressure can be maintained in this way.

Such an active pressure control is not required by the IGF-Code. Instead, the idea of homogenous liquid expansion (comp. Sect. 12.1.2, p. 102) is the basis of the safety philosophy. In addition, the assumption that a severe fire can lead to unacceptable liquid expansion, is used to weaken the LL requirement of [3], Sect. 6.8.1 for type C tanks. Sect. 6.8.2 allows a maximum filling level of 95% for "...cases where the tank insulation and tank location make a probability very small for tank content to be heated up due to an external fire".

For the author, the 95% is simply a guess. Apart from the claim from designers that the calculated LL, well below 90% for type C tanks, is too low and an unfair regulation a justification for the 95% is not known to the author. The author partly agree to the designer's view but there should be a physical justification for the maximum LL.

12.1.4.2   Liquid Expansion Under Fire Conditions

The fire conditions assumed for PRV sizing are explained in Sect. 7.2, p. 64. Evidence has been provided by extensive $R \& D$ work in the 1980s and early 1990s that a fire may lead to a liquid full type C tank if 98% LL is used [5, 6] but that the liquid protects the tank and delays a tank failure by several hours (comp. Figure 12.6). The $R \& D$ work also confirmed that PRVs sized according the IGC-Code, IGF-Code are adequate to protect a liquid full tank with 2-Phase flow through the valves.

---

[11] To the knowledge of the author the 15 d come from older regulations in the US.

[12] E.g. in case of a grounding or a collision.

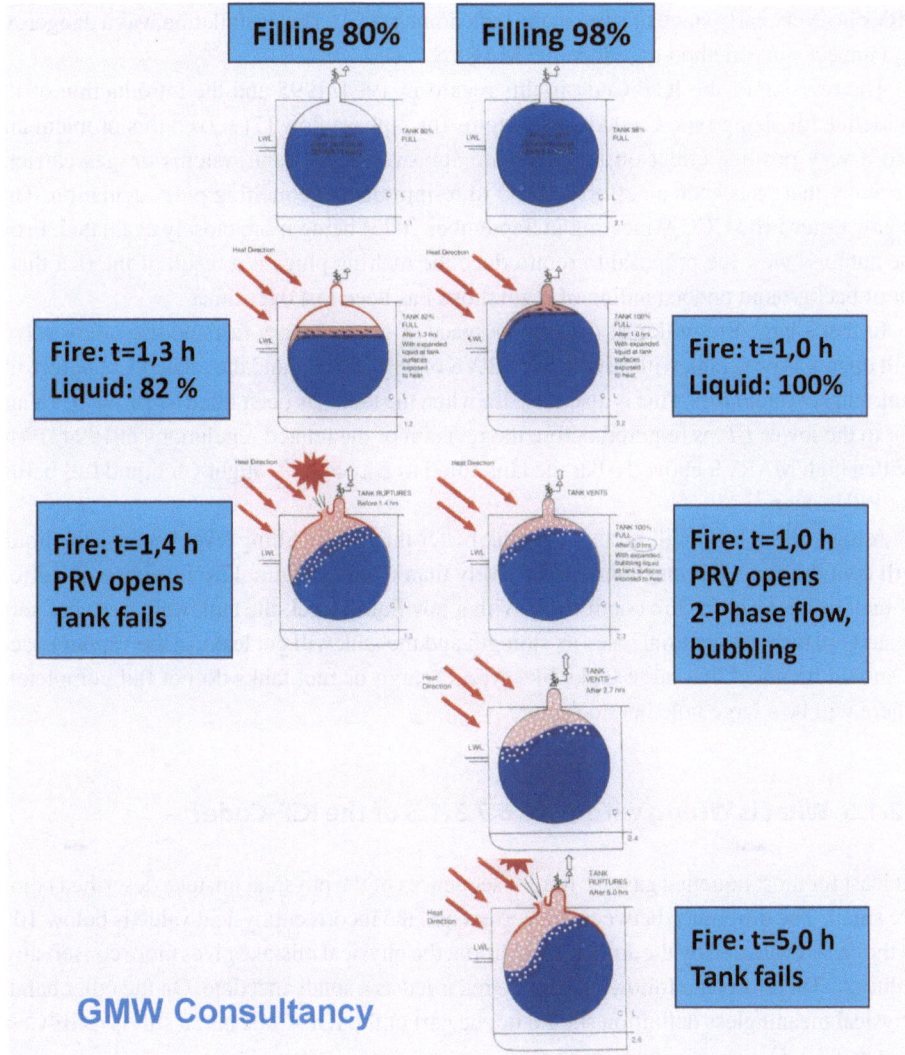

Fig. 12.6 Illustration of fire effects and tank filling. (*Source* GMW Consultancy, comp. also [5])

To avoid low filling limits related to a high design pressure of a type C tank the PRVs had been arranged by the installation of melting plugs to open at a low MARVS[13] in case of fire because the LL was related to the MARVS only. The *R & D* work named above was done in connection with the revision of the filling limit requirements for type C tanks which aimed and achieved to delete the need to install melting plugs close to the PRVs which activate the

---

[13] To 250 mbar g in most cases.

PRVs in a very early stage of a fire in the tank dome area.[14] This installation was a dangerous and unnecessary method to reduce the MARVS.

The revision of the IGC-Code in this regard in 1994/1995 and the introduction of the guideline for sizing type C tank vent systems for 2-phase flow [7] solved this problem and had a very positive effect on the adequate approval of all vent systems on gas carriers. Recently there has been an effort at IMO to re-introduce the melting plug regulation. This is now (after IMO CCC-9 meeting in September 2023) being more closely evaluated. From the author's view the proposal to reintroduce the melting plugs is a result of the fact that a lot of background understanding of regulations has been lost over time.

In fact a large fire will heat the vapour space of the tank very fast and the safety valves will open a Type C tank with a typical MARVS of, e.g., 5 bar g and the related LL, before the tank can get liquid full. This is also the case when the tank has been filled to $LL = 0, 98$ and not to the lower LL as required before the revision of the related regulations in 1994/1995. With a high MARVS above 10 bar the tank filled to $LL = 0, 98$ might get liquid full before the PRVs open.[15]

A high filling level will protect the tank better than a low filling level because the liquid will cool the tank wall much more effectively than the gas. Figure 12.6 illustrates the effect of tank filling level for fire conditions. With a low liquid level, the tank walls are very soon heated up, the tank material loses its strength and the tank will get leaks in the vapour space. It should be noted that large tanks like type C cargo or fuel tanks do not fail completely. There will be a large hole but no rupture.

12.1.5 What is Wrong with Sect. 6.7.3.1.3 of the IGF-Code?

At least for most liquefied gases[16] the consequences of the physical mistake described below are small. The difference between the correct and the incorrect physical value is below 10% in the cases evaluated by the author. In addition, the physical mistake gives more conservative values.[17] Therefore, the following may be regarded as a small anecdote. On the other hand a physical meaningless definition should not be part of the IGF-Code but, it still is (IGF-Code Sect. 6.7.3.1.3).

The physical background of PRV sizing and the pressure loss calculation in IGC-Code and IGF-Code [1, 3] may be is difficult to understand but it is existing. For this reason, it is possible to prove the sizing on the correct physical base, as recommended in Sect. 7.3, p. 67. But a meaningless definition, as give in IGF-Code Sect. 6.7.3.1.3, is a problem. The author regards this as a serious mistake.

---

[14] Even in case of a small local fire close to the PRVs.

[15] Note that the MARVS used for the illustration of Fig. 12.6 had this high MARVS.

[16] The author has not checked all 37 gases listed in IGC-Code [1] plus Hydrogen.

[17] Which means higher pressure loss with related bigger piping.

As explained the PRV sizing for the fire case in industry and shipping are based on a fire load and the assumption that the heat flux from the fire load leads to the evaporation of the tank content [5]. The vapour created is released by the PRVs.

The IGC-Code and IGF-Code as well as US refinery and transport rules[18] on this subject use the same physical background.[19] From this a theoretical volume flow of air at ISO standard conditions for air[20] (Q comp. Sect. 6.7.3.1.1 IGC-Code) is used for sizing the valves according IGC-Code, IGF-Code. This air volume flow is based on the physical properties of the liquefied gas transported (IGC-Code) or fuel (IGF-Code). Therefore, the volume flow of "free air" can be traced back to the named assumptions and the heat flux in a physical correct way. It must be admitted that this is not obvious from the formulas and explanation in the rules.

The introduction of requirements for pressure loss calculations by the IMO resolution A.829(19) in 1995 [7] was a real positive step to improve the vent lines upstream and downstream of PRVs. The guideline of this resolution includes a procedure for single gas-phase flow and for 2-phase flow. Meanwhile, the procedure is widely used in shipping and reflected by class rules. The point is that the pressure loss calculation can not be made with the sizing volume flow at ambient temperature and pressure conditions named above. Therefore, a mass flow is needed for the pressure loss calculation.

This mass flow is created by the assumed fire load used for the PRV sizing volume flow.

$$\dot{Q}_{fi} = 71 \cdot 10^3 \cdot F \cdot A^{0.82} = \dot{W}_g \cdot h_{fg}; \quad [kW = (kg/s) \cdot (KJ/kg)] \tag{12.1}$$

In Eq. 12.1 the mass flow of evaporated cargo (IGC-Code) or fuel (IGF-Code) **must** be used! To calculate this, the IGC-Code, IGF-Code use a specific heat flux of 71 kW/m^2, a tank area effected by the fire of $A^{0.82}$ and a fire factor F.[21]

All this only makes sense if the mass flow of gas (W_g) of cargo/fuel at relieving conditions together with the heat of evaporation (h_{fg}) of cargo/fuel is used[22]! The pressure loss calculation **must** be done by using W_g. This is the case for gas carriers because the calculation follows resolution A.829(19) [7]. This is **not** done for gas fuelled ships because the mass flow for the pressure loss calculation for IGF-Code is given in Sect. 6.7.3.1.3. This section states.

"The required mass flow of air at relieving conditions is given by:

$$M_{air} = Q \cdot \rho_{air} \tag{12.2}$$

where density of air (ρ_{air}) = 1, 293 kg/m^3 (air at 273, 15 K, 0, 1013 MPa)."

---

[18] Rules of API, CGA: IGC-Code [1] Sect. 8.4, IGF-Code [3] Sect. 6.7.3 are based on these rules.

[19] Detailed evaluation comp. Annex C, p. 137.

[20] 1, 013 bar, 273, 15 K

[21] For details please compare the IMO resolution A.829(19) [7], Sec "3. Equations".

[22] Comp. Eq. 1. in resolution A.829(19) [7].

The value of Q in this equation is the volume flow of free air according Sect. 6.3.1.1 of the IGF-Code. This requirement has no physical justification and is simply wrong.

> The author worked actively in IMO until 2013 as consultant for the German Ministry of Transport and with SIGTTO (SIGTTO: Society of International Gas Tanker and Terminal Operators) on IGF-Code development and IGC-Code matters including the revision of the IGC-Code 2008/2010. The colleagues R. Gray, D. Chatburn (SIGTTO), B.O. Bauer Nilsen (DNV), G. Wuersig (GL) developed the IMO Resolution A.829(19) in 1992/1995.
>
> As explained, the wrong interpretation in Sect. 6.7.3.1.3 is today part of the current IGF-Code [3]. The correct one is part of the IGC-Code [1] because the Code makes reference to IMO Resolution A.829(19).

12.1.6 Conclusions for Tank Loading Limits

From the above evaluation the author's conclusions for tank filling limit definition of liquefied gas tanks in shipping are:

- Liquefied gases will not expand if the pressure is kept constant by use of BOG. Therefore, high filling limits can be accepted. Anyhow, uncertainties in liquid expansion and PRV opening by sloshing should be avoided.
  The author's view is that a maximum LL of 98% should not be exceeded in liquefied gas transport and fuel application.
- For gas carriers the current regulations for loading limit (LL) as definition in Sect. 15.5 of the IGC-Code [1] should be kept as the standard procedure. They should be amended by the requirements for the heat balance calculation named below.
- For gas fuelled ships the current regulations for loading limit (LL) definition in Sect. 6.8.1 of the IGF-Code [3] should be kept as a standard procedure. They should be amended by the requirements for the heat balance calculation named below. The requirement according to Sect. 6.8.2 is not necessary anymore if the named calculation is included.
- The assumption of homogeneous energy distribution in the tank content is not correct in most cases. Therefore, in most cases an active pressure control is needed for for all tank types if the use of gas from the tanks can not be assured. E.g., in case of a ship incident or long term system breakdown. A fixed filling level [23] above the value which is equal to a filling below 98% of tank volume for saturation conditions at MARVS should not have been stipulated without considering the pressure and temperature control.

---

[23] E.g. 95% as in IGF-Code Sect. 6.8.2.

- The LL definition should include the calculation of a heat balance to ensure that the tank can not get liquid full and the pressure can not reach MARVS within a given limit in time.

  For this calculation a full tank and no consumption beside the consumption related to the control of tank pressure and temperature should be assumed. This time should be included in the IGC-Code for gas transport and used for all liquefied gases. The current definition for ships using LNG as fuel of 15 d in Sect. 6.9.1.1 [3] is regarded by the author as a reasonable value.

References

1. IMO (2016), IMO Resolution MSC.5(48), MSC.370(93), IMO IGC-Code, International Code for the Construction and Equipment of Ships Carrying Liquefied Gases in Bulk, IMO London, ISBN 978-92-801-1631-1
2. P. Hauschildt, Auswertung von Versuchen zum thermodynamischen Verhalten von verflüssigtem Gas an einem 61 m^3 Flüssig-Wasserstoff Behälter; Diplomarbeit, Technische Universität Hamburg Harburg, Hamburg Dez. 1997
3. IMO (2016), IMO Resolution MSC.391(95), IGF-Code: International Code of Safety for Ships using Gases or other Low-Flashpoint Fuels, IMO, London, ISBN 978-92-801-1653-3
4. US Code of Federal Regulations, Dated: 20-09-2023, title-46: shipping, chapter-I: coast Guard, Department of Homeland Security, Part-154: Safety Standards for Self-Propelled Vessels Carrying Bulk Liquefied Gases, Subpart O: Certain Bulk Dangerous Cargoes, Subpart C-Design, Construction and Equipment, No-703: Methane(LNG)
5. M. Böckenhauer, G.M. Würsig (GL), R.H. Chadburn (SIGTTO), B.O. Bauer-Nilsen (DNV),The New Cargo Tank Loading Limit Requirements in the IMO Gas Carrier Codes, GASTECH-1994 Conference proceedings, Kuala Lumpur, 1994
6. REVIEW OF THE IGC CODE - Information on the development of IGC Code chapter 15, Submitted by SIGTTO, CCC 9/INF.24, IMO, London, 18 July 2023
7. Guidelines for the Evaluation of the Adequacy of Type-C Tank Vent Systems, IMO Assembly Resolution A.829(19), adopted on 23rd November 1995, IMO, London

Fire on Board Ships

13

The major difference between a fire or explosion event on board of a ship and on shore, e.g., when a car is burning on a motorway is, that it is not a possibility to run away on board of a ship.

Fire and explosion is a relevant incident cause on board ships. According to the Allianz incident report 2023 [1], about 7% of all incidents in 2022 were related to fire and explosion events.[1] Figure 13.1 gives an overview of the causes of incident and losses.

Figure 13.1 show that fire and explosion events are in the same range as collisions between ships and stranding/grounding events.

Depending on the fire load, ship fires can be very long lasting especially if the cargo is the fire load. The different parts of the ship are separated terms of fire mitigation by structural measures. Incidents like the fire on board of the RoRo ship Lisco Gloria (2010) or the car carrier Fremantel Highway (2023) illustrate that structural separation is obviously working to limit the spreading of ship fires.

After the long lasting fires, the areas with intact ship's hull paint is intact indicate in both cases that the fires in these large areas were less sever than in the direct neighbourhood. In fact, it has been reported in the press that the cars on the lower car decks of the MS Fremental Highway were in good condition. At the same time the cars in the neighbouring compartments were destroyed completely. The main aim of the structural fire protection is to limit the extent of the fire and to "buy time" for firefighting and in the worst case for evacuation.

---

[1] 209 cases out of 3032.

G. Würsig, *The Safety Principles for the Use of Low Flashpoint Fuels in Shipping*,
Synthesis Lectures on Ocean Systems Engineering,
https://doi.org/10.1007/978-3-031-64174-9_13

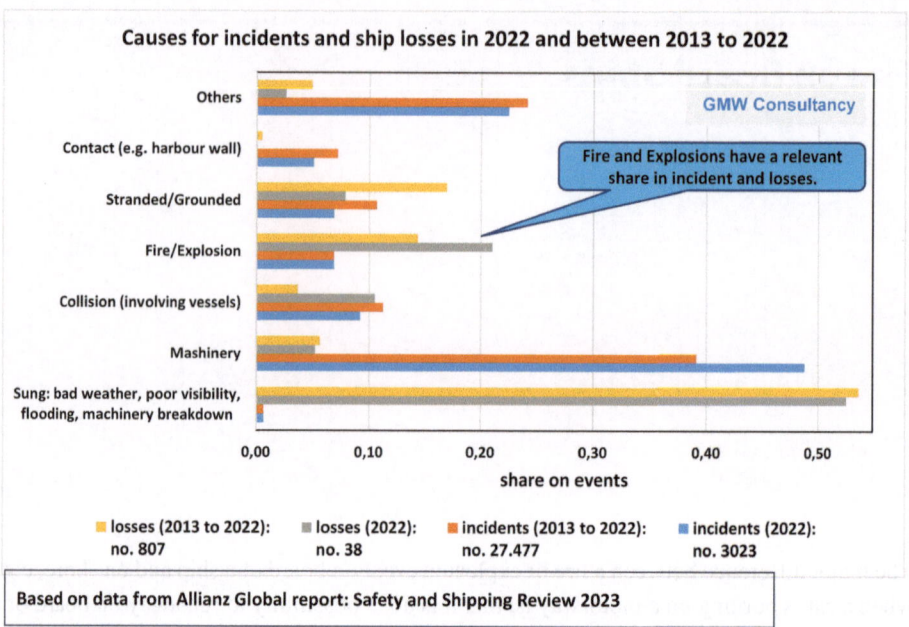

Fig. 13.1 Causes for incidents on ships and ship losses in 2022 and in the time frame 2013 to 2022 compiled from Allianz Global report [1]. (*Source* GMW Consultancy)

For ship fuel systems the aim in terms of the risk of fire and explosion is to limit the consequences to the ambient in the case of a fire/explosion caused by the ship's fuel. At the same time the spreading of external fire and explosion events needs to be mitigated.

13.1   Limitation of Fire Loads from Cargo and Fuels

The examples above illustrate that a tank with alternative fuel might be subject to long lasting high fire loads from cargo fires. The good news is that the PRVs are sized for such fires. If the fuel tank is located in a hold, it is likely that the liquid in the tank will avoid a tank failure for quite a long time. For an external fire of cargo, the fuel tank installation below or partly below the waterline is the best protection. Of course, such a location subjects the fuel tank to a higher risk in case of collision. For the author the regulations of IGF-Code Sect. 5.3.4 are the best compromise for mitigation of the different factors if they are applied in the correct way (for this, please comp. Sect. 8.2, p. 74)

For tanks above deck the fire load from cargo is more relevant than for tanks below deck. In a lot of cases such tanks are installed on tankers and bulkers. Possibly with the exception of bulkers for non flammable cargo[2] the potential fire load of the cargo itself is often the much higher fire load compared to the fuel. From experiences with gas carriers it can be concluded that the current level of protection for such deck tanks is an adequate mitigation measure.

Compared to the potential fire load of the fuel stored in the fuel tanks, the fire load of the fuel system downstream of the fuel tanks is relatively low and should be kept low by design. E.g., the volume of fuel buffer tanks should be as low as possible and the total number of all components containing high amounts of fuels like heat exchangers should be limited to the needed minimum. This requirement does not mean that all piping must be purged and inerted in any case of system stops! Purging and inerting should be ensured if a stop of the system containing alternative fuel is for a longer duration or if the duration cannot be limited because it is not known. In most cases the pressure reduction in pressurized systems and the removal of cold liquefied gas from system components is sufficient to bring the systems into a safe condition. Note that the current IGF-Code "sees" this different in some cases.

Besides the fuel storage tanks, the regular bunker operations present the highest potential fire loads on ships with alternative fuels. Specially adequate couplings (DQRDC type) are a must have safety measure even for large fuel tanks (comp. Chap. 11, p. 91). Short length of hoses, avoidance of hose couplings and double walled hoses with leak detection are strongly recommended. The author's view is that this should become a requirement at least for ships with a large number of third party persons on board and in the ambient near the bunker operations. In other words passengers, on ferries and cruise ships.

Reference

1. Allianz Global-Corporate and Specialty, Safety and Shipping; Review 2023 An annual review of trends and developments in shipping losses and safety, www.agcs.allianz.com

---

[2] E.g. stones.

Explosion Protection

The differences and similarities between explosion deflagration and deflagration are discussed in Sect. 4.3, p. 26. The following is limited to deflagration events and the term explosion is used as a synonym.

14.1 Ventilation

Considering the fact that there are no tight systems in the world (comp. Sect. 3.3, p. 20) natural and forced ventilation are most important when designing safe and reliable alternative fuel systems. The aim of the ventilation is twofold:

- The normal operation of the system should not be disturbed by unavoidable minor leakages. For this reason, the ventilation should be high enough to remove traces of fuel gas from the system without creating an alarm or shut down.
- At the same time the ventilation should be high enough to keep the atmosphere in a space below the lower explosion limit for a sufficient time to mitigate a release by e.g. a shut down. This concept is, e.g., applied for ESD engine rooms.

For enclosed spaces these aims are achieved by forced ventilation and for open spaces by natural air exchange. IGF-Code [1] gives the related requirements in Sect. 13.

In shipping, as in other industries, an air exchange rate of 30 times the empty room volume per hour is regarded to be a sufficient ventilation rate to achieve the aims given above. To ensure normal operation also reduced rates, to 5 or 10 times the empty room volume per hour, are sometimes used. In all cases the volume flow is increased to 30 air changes in case

G. Würsig, *The Safety Principles for the Use of Low Flashpoint Fuels in Shipping*,
Synthesis Lectures on Ocean Systems Engineering,
https://doi.org/10.1007/978-3-031-64174-9_14

of leak detection. For comparison, the natural air exchange rate of a living room is approx. 1 room volume per hour.[1]

By the way, the author was not able so far to get a real answer why 30 is the magic exchange rate? In practice 30 air changes are relatively high for large rooms but an easy to reach value which do not create "a storm" in the room. The author's impression is that the logic behind this value has been lost over time. Assuming this, the author hopes that some reasoning behind alternative fuel safety requirements may survive with this book. At least for some time.

14.2   Ex-Zoning

The idea of Ex-Zone requirements for technical equipment to the best of the author's knowledge was first introduced for electrical equipment. Today it has been expanded to other types of equipment which might create an ignition source, but electrical equipment is still in focus. The most recognized standard to define Ex-Zones for areas where flammable gas might be present is IEC 600-79-10-1 [2]. Perhaps because shipping loves prescriptive regulations, the evaluation of Ex-Zones according IEC 60079-10-1 is only permitted in exceptional cases.[2] Consequently IGF-Code Sect. 12.5 defines the related areas.

The commonly used definition for EX-Z has been developed by the INTERNATIONAL ECTROTECHNICAL COMMISSION (IEC) and is given e.g. in IEC-60079-10-1 which is the relevant standard for the interpretation of Ex-Zones in IGF-Code and IGC-Code [2]. IGC-Code gives the definition in Sect. 10.1. It has to be noted, that the definition apply to any equipment and is not limited to electrical equipment. To highlight electrical equipment is related to history because the IEC developed the concept and the requirements.

The IGF-Code does not give a dedicated definition. It was included in Sect. 1.3.12 of the IMO interim guideline for gas fuelled ships in 2009 [3] but for whatever reason not in the Code from 2015 [4]. Not even a reference to IGC-Code definition is given. According to IGC-Code, Sect. 10.1 the zones are defined as follows:

- "Zone 0 ...is an area in which an explosive gas atmosphere **is present continuously** or is present for long periods."
- "Zone 1 ....is an area in which an explosive gas atmosphere **is likely to occur** in normal operation."
- "Zone 2 ... is an area in which an explosive gas atmosphere **is not likely to occur** in normal operation and, if it does occur, is likely to do so infrequently and for a short period only."

---

[1] If not designed according to the latest German insulation and tightness requirements for buildings. These buildings require a forced ventilation to ensure an adequate air exchange.
[2] E.g., for fuel cell applications.

- "Non-hazardous area is an area in which an explosive gas atmosphere **is not expected** to be present in quantities such as to require special precautions for the construction, installation and use of electrical apparatus."

From physics, an "explosive gas atmosphere" requires the presence of burnable gas and oxygen. The only thing missing to create an ignition is an ignition source. For tanks this condition is fulfilled if the tank has an opening to air as it is the case for oil fuel tanks or if inerting is not required. Tanks for liquefied gases are purged, inerted and then filled with the liquefied gas. Regardless of this, IGF-Code Sect. 12.5.1 define them as zone-0.

Also, the definitions of other room types in IGF-Code Sect. 12.5.2 as zone-1 is on the conservative side. Considering the ventilation type and design, most of these rooms would be zone 2 according IEC-60079-10-1.

A rule is a rule and exceptions are not permitted!
If it is stipulated that a number of cascading operational failures occur, accumulation of air in a LNG tank might be possible even if it must be assumed that the water vapour in the air would be frozen. This would not be a normal operation any more. Accumulation of air in a tank with LH2 is definitely not possible because not only the water vapour but also the oxygen, nitrogen will be frozen and in the best case accumulate at the tank bottom. For this reason, a zone 0 with permanent explosive atmosphere is physically impossible for an LH2 tank.

When the author was responsible for the tests with the 61 m³ LH2 tank illustrated in Fig. 12.5, p. 104, he also was responsible for the installation of more than 30 temperature sensors in the gas and liquid space of the tank.

Approval was done by his colleagues in Germanischer Lloyd. They insisted that the sensors must be approved and certified for Zone 0 operation. Such sensors were not existing. The only possible way to get the approval for the temperature equipment was to reduce the electrical energy for getting the sensor signal below the minimum ignition energy of hydrogen/air mixtures at ambient conditions!

At the end it was necessary to read out one sensor signal after the other which needed more than a minute for all sensors. The original idea was to read out the signals of all sensors at 10–100 Hz. A lot of dynamic data was lost by using the speed of 0, 017 Hz instead of 10–100 Hz.

14.2.1 Protection Against Overpressure

In the event of a deflagration, the created overpressure must be relieved to avoid the destruction of the compartment where the event occurs. The most critical components for such a deflagration are:

- Suction and exhaust systems for engines. Adequate protection can be done by the use of burst discs. Example given in Fig. 14.1.
- ESD engine or other ESD rooms. For these rooms it is possible to simply install "weak" wall parts which definitely fail long before the room walls are affected. E.g., an aluminium plate of adequate size screwed to a wall opening is a simple maintenance free solution. Note that the direction of possible release must consider that personnel can not be injured in case of operation.
- Crankcases of engines. Engine crankcases are protected against internal explosions by explosion flaps to protect the crankcase from oil mist explosions which might be caused by hot internal surfaces.

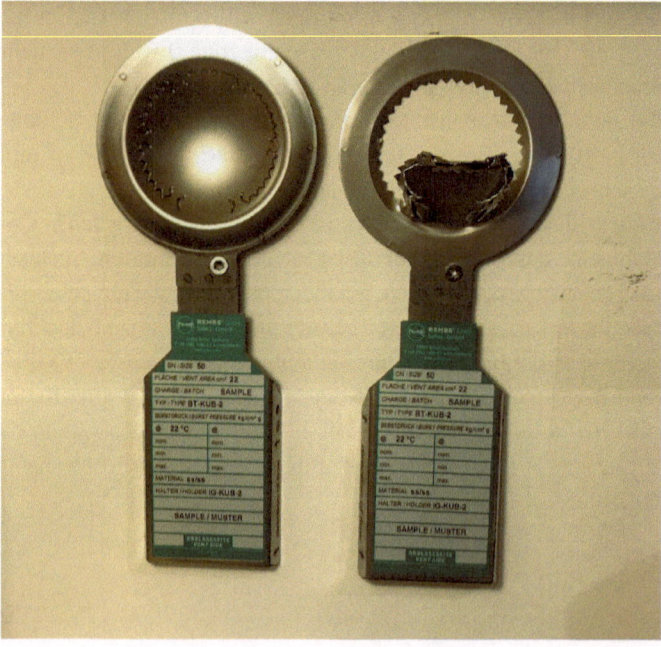

Fig. 14.1 Examples for a burst disc before and after operation. (*Source Photo* Dr. Gerd Wuersig, Burst disk: REMBE GmbH Safety+Control, Gallbergweg 21, 59929 Brilon, Germany)

At least if LNG is used as fuel, the fuel does not create an additional risk in engine crankcases because the oil mist will ignite much before the Methane gas. For this reason, the requirement for gas detection in crankcases undermines the safety culture. It is unlikely that safety culture is improved by frequently occurring faulty alarms.

For the above measures, it must be ensured that the room around the protected equipment is large enough or has the necessary permanent openings to avoid an unacceptable overpressure.

References

1. IMO (2016), IMO Resolution MSC.391(95), IGF-Code: International Code of Safety for Ships using Gases or other Low-Flashpoint Fuels, IMO, London, ISBN 978-92-801-1653-3
2. Explosive atmospheres–Part 10-1: Classification of areas–Explosive gas atmospheres; IEC Standard 60079-10-1, 2021,INTERNATIONAL ELECTROTECHNICAL COMMISSION (IEC), ISBN 978-2-8322-9213-6
3. IMO RESOLUTION MSC.285(86), adopted on 1 June 2009, INTERIM GUIDELINES ON SAFETY FOR NATURAL GAS-FUELLED ENGINE INSTALLATIONS IN SHIPS, London, 2009
4. IMO (2015); RESOLUTION MSC.391(95) (adopted on 11 June 2015) ADOPTION OF THE INTERNATIONAL CODE OF SAFETY FOR SHIPS USING GASES OR OTHER LOW-FLASHPOINT FUELS (IGF CODE), London

Final Remark

15

The aim of this book is to motivate the experts in design, application, and rule setting for alternative fuels in shipping to consider the background of their work in a logical and reasonable way.

Nearly all design principles in shipping, except the design of the ship's hull itself, are not unique to shipping. In fact, to a very large extend shipping is applying what is used in much larger quantities in other industries. However, some things are more critical on board of ships. E.g., it is not possible to run away in case of a fire and it is not a good idea to switch off power for safety reasons during a storm. Considering that shipping may be special in some cases but not in general, this book may be also interesting for engineers in other industries namely in chemical and refinery industry. At least the author, as a mechanical, process engineer and a member of the process safety working group at DECHEMA,[1] maintains this hope.

Knowing that rules and regulations hide their background in a very effective way and that this background can be lost over time, the author is aiming to prevent at least some ideas behind the current regulations for alternative fuelled ships from getting lost. At least for some time.

This book is not a textbook. It also is not a novel to be read from the first to the last page. The reader may choose his topics and follow the different links which are included for a better understanding and to avoid duplications. This book is not intended to be a learning aid therefore e.g., the glossary reflects the explanations looking relevant for the author and is not the "100% version".

The author tries to explain the reasoning and background of some rules and regulations in the IGF-Code [1], IGF-Code [2] and to give hints their reasonable application, weak points

---

[1] DECHEMA: "Deutsche Gesellschaft für chemisches Apparatewesen (German Society for Chemical Apparatus")".

and areas for improvement. Following the principle that life is too serious to take it serious, he has not hesitated to include some personal remarks and anecdotes from his professional life.

References

1. IMO (2016), IMO Resolution MSC.391(95), IGF-Code: International Code of Safety for Ships using Gases or other Low-Flashpoint Fuels, IMO, London, ISBN 978-92-801-1653-3
2. IMO (2016), IMO Resolution MSC.5(48), MSC.370(93), IMO IGC-Code, International Code for the Construction and Equipment of Ships Carrying Liquefied Gases in Bulk, IMO London, ISBN 978-92-801-1631-1

Appendix
Carnival Corporation Finalizes Contract with Meyer Werft

G. Würsig, *The Safety Principles for the Use of Low Flashpoint Fuels in Shipping*,
Synthesis Lectures on Ocean Systems Engineering,
https://doi.org/10.1007/978-3-031-64174-9

DE | EN    **MEYER WERFT**
PAPENBURG 1795

Press › Press Detail

15.06.2015

Carnival Corporation Finalizes Contract with Meyer Werft to Build Four Next-Generation Cruise Ships

BACK TO OVERVIEW

The four new cruise ships – part of a previously announced nine-ship strategic partnership for the world's largest cruise company – will be the largest ever built based on guest capacity

The vessels will feature a revolutionary "green cruising" design as the first-ever cruise ships powered at sea by Liquefied Natural Gas, the world's cleanest burning fossil fuel

MIAMI, June 15, 2015 — Carnival Corporation & plc, the world's largest travel and leisure company, today announced it has signed a multi-billion

dollar contract to build four next-generation cruise ships with the largest guest capacity in the world. The contract with Meyer Werft is part of larger previously announced strategic memo of understanding with leading shipbuilders Meyer Werft and Fincantieri S.p.A for nine new ship orders between 2019 and 2022.

The four new ships will also feature a revolutionary "green cruising" design. The ships will be the first in the cruise industry to be powered at sea by Liquefied Natural Gas (LNG) -- the world's cleanest burning fossil fuel, representing a major environmental breakthrough.

The company said two of the ships will be manufactured for AIDA Cruises at Meyer Werft's shipyard in Papenburg, Germany. Additional information about the ships, including which new ships will be added to each brand, will be made available at a later date.

Based on Carnival Corporation's innovative new ship design, each of the four next-generation ships will have a total capacity of 6,600 guests, feature more than 5,000 lower berths, exceed 180,000 gross tons and incorporate an extensive number of guest-friendly features. A major part of the innovative design involves making much more efficient use of the ship's spaces, creating an enhanced onboard experience for guests.

Pioneering a new era in the use of sustainable

fuels, the four new ships will be the first in the cruise industry to use LNG in dual-powered hybrid engines to power the ship both in port and on the open sea. LNG will be stored onboard and used to generate 100 percent power at sea – producing another industry-first innovation for Carnival Corporation and its brands. Using LNG to power the ships in port and at sea will eliminate emissions of soot particles and sulfur oxides.

In addition to the two ships being built in Germany, Meyer Werft – which had the capacity to accommodate these four ship-building orders in its production schedule -- will also build the two additional ships detailed in today's announcement at its shipyard in Turku, Finland. Each new ship will be specifically designed and developed for the brand and the guests it will serve, underscoring the company's goal to consistently exceed guest expectations and provide first-time and repeat guests with the vacation experience of a lifetime on each and every cruise.

Carnival Corporation CEO Arnold Donald said the contract is consistent with the company's measured capacity growth strategy to replace ships with less efficient capacity with newer, larger and more fuel efficient vessels over time.

"We are looking forward to executing on the next step in our fleet enhancement plan," said Donald. "At a cost per berth in line with our existing order book, these new ships will enhance the return profile of our fleet. These are exceptionally efficient ships with incredible cabins and public spaces featuring a design inspired by Micky Arison and Michael Thamm and developed by our new build teams." Arison is chairman of the board of directors for Carnival Corporation & plc and Thamm is CEO of the Costa Group, which includes AIDA Cruises and Costa Cruises.

Added Donald: "It will be exciting to see our shipbuilding team bring these new ships to life. Every step of the way, our focus is on designing state-of-the-art ships that provide a vacation experience our guests will love, and we are putting all of our creative energy and resources into making sure we achieve that goal."

"These ships will expand our leadership position for the Costa Group, the market leader in all the major European markets," said Thamm. "These will be spectacular ships designed specifically for our guests who sail on our Costa Group brands."

Bernard Meyer, CEO of Meyer Werft, said: "In past years, we have built seven highly successful ships for AIDA Cruises. We are honored that Carnival Corporation has entrusted us with the implementation of this ambitious shipbuilding program, and we look forward to building these four magnificent ships."

The new ship order will allow the Costa Group to continue to build on its leadership position in the

European cruise market – a market in which five out of 10 cruise guests in 2014 sailed onboard a Costa Group ship. The Costa Group – along with Princess Cruise Lines, also part of the Carnival Corporation family -- also occupies the leading position in the rapidly growing cruise market in China.

As part of each shipbuilding company's long-term strategic partnership with Carnival Corporation, additional new ship orders are being explored over the coming decade.

About Carnival Corporation & plc

Carnival Corporation & plc is the largest cruise company in the world, with a portfolio of 10 cruise brands in North America, Europe, Australia and Asia, comprised of Carnival Cruise Lines, Holland America Line, Princess Cruises, Seabourn, AIDA Cruises, Costa Cruises, Cunard, P&O Cruises (Australia), P&O Cruises (UK) and fathom.

Together, these brands will operate 100 ships in 2015 totaling 219,000 lower berths with eight new ships scheduled to be delivered between 2016 and 2018, along with an additional four ships to be delivered between 2019-2022. Carnival Corporation & plc also operates Holland America Princess Alaska Tours, the leading tour companies in Alaska and the Canadian Yukon. Traded on both the New York and London Stock Exchanges, Carnival Corporation & plc is the only group in the world to be included in both the S&P 500 and the FTSE 100 indices.

Share

search

COMPANY   VISITORS

TECHNOLOGIES   PRESS

SHIPS   SUPPLIERS

CAREER

 FOLLOW MEYER WERFT

What happens if LH2 is transported in LNG tanks?
GMW Consultancy: Annex

Dr. Ing. Gerd Würsig - GMW Consultancy -

January 8, 2024

G. Würsig, *The Safety Principles for the Use of Low Flashpoint Fuels in Shipping*,
Synthesis Lectures on Ocean Systems Engineering,
https://doi.org/10.1007/978-3-031-64174-9

For liquefied gas tanks on gas carriers and gas fuelled ships the heat flux into the tanks is generally expressed by the BOR in % of gas evaporated per day. The gas is called BOG. The BOR is a common way to express the heat flux from the ambient into the liquefied gas tank. It is defined by

$$\dot{Q}_{fg} = \dot{M} \cdot (h'' - h') \ \ [W] \tag{1}$$

$$\dot{q}_{fg} = \dot{Q}_{fg}/A_{Ta} \ \ [W/m^2] \tag{2}$$

Typical BOR rates for LNG tanks are $0,12 \ \%/d$ to $0,20 \ \%/d$ for large tanks of LNG carriers and $0,3 \ \%/d$ to $0,60 \ \%/d$ for LNG fuel tanks. For a large Moss tank Eq.1, 2 is as follows.

With the inner diameter of the sphere ($d_i = 44 \ m$) the tank surface become:

$$A_{Ta} = 6.050, - \ [m^2]$$

The heat of evaporation of the LNG at $1,250 \ bar \ abs$ is

$$(h'' - h') = 800 - 295 = 505 \ \ [kJ/kg)]^1$$

Assuming a filling level of $0,98$ [2]

$$M = 0,98 \cdot V_0 \cdot \rho_{fl} = 0,98 \ \cdot 44.250 \ [m^3] \cdot 422 \ [kg/m^3] = 18.300.030, -[kg]$$

The mass of BOG at an assumed low BOR of $0,13 \ \%/d$ is:

$$BOG = 18.300.030, -[kg] \cdot 0,12/100 \approx 23.800, - \ [kg/d]$$

The resulting heat flux to generate this BOG is

$$\dot{Q} = 23.800 \cdot 505 \ [kJ/kg] \cdot \frac{1}{24 \cdot 3600} = 139 \ [kW]$$

If the share of heat flux through the tank supports and piping is assumed with 20 % the remaining heat flux through the insulation in this illustrating example is $Q_{is} = 0,8 \ \cdot 139 = 111 \ [kW]$ or $q_{is} = 18, -[W/m^2]$. Using a temperature difference between LNG and ambient of $180 \ K$, an insulation thickness of $s_{is} = 400 \ [mm]$ and a simplified approach for the heat transfer by assuming $k = \lambda_{is}/s_{is}$ gives a thermal conductivity of the insulation material of

$$\lambda_{is} = \frac{q_{is} \cdot s_{is}}{\Delta T_{is,ab}} = 18 \cdot 0,400 \cdot 1/180 = 0,04 \ [W/(m \cdot K)]$$

Which is in the range of typical polyurethane based tank insulation systems.

Is it possible to transport LH2 in this tank?

Assuming that the problem of condensing Nitrogen and Oxygen from air at the outer tank wall of the LH2 tank is solved and all tank materials are suitable for LH2 the above calculation can be done for LH2.

---

[1] her and in the following properties of Methane are assumed for the LNG
[2] sec 15.4 of IGC-Code gives a mass of LNG

- GMW Consultancy - Annex 2

With LH2 at a temperature of 21 K $(1,25\ bar\ abs)$ and a density of 71 kg/m^3 but for the same tank gives a BOR of $1,75\%/d$. This is $13,5$ times the LNG BOR. Obviously this is to high for a reasonable business case of LH2 ship transport.[3]

The simple increase in insulation thickness from 400 mm by the factor of $13,5$ to $5.400\ mm$ is only a part of the solution because the heat transfer through the bearings would lead to a BOR of still $0,9\%(d)$. Anyhow, there are more clever ways to design LH2 carrier tank systems with reasonable BOR.

From the above calculations it is concluded that a transport of LH2 with existing LNG carriers is not possible and a simple adaption of existing LNG designs is not the way forward.

---

[3]Note: From published technical data for the LH2 carrier Susio Frontier the author calculated a BOR of more than $1,3\ \%$ per day.

Appendix   **C**

The Relations Between the Rules for Sizing PRVs for the Fire Case

Comparison of API, CGA, IGC-Code and IGF-Code
requirements for PRV sizing for fire conditions

Dr. Ing. Gerd Würsig - GMW Consultancy -

April 15, 2024

© The Editor(s) (if applicable) and The Author(s), under exclusive license to Springer   137
Nature Switzerland AG 2025
G. Würsig, *The Safety Principles for the Use of Low Flashpoint Fuels in Shipping*,
Synthesis Lectures on Ocean Systems Engineering,
https://doi.org/10.1007/978-3-031-64174-9

Contents

Chapter 1

The API, CGA, IGC-Code and IGF-Code regulations

Historically the most common regulations used for dimensioning of safety devices under fire conditions in shore site industry are the API and CGA regulations. This report give the background of the regulations which are also used for the International Code for the Construction and Equipment of Ships Carrying Liquefied Gases in Bulk (IGC-Code) and the International Code of Safety for Ships using Gases or other Low-Flashpoint Fuels (IGF-Code). With regard to Pressure Relief Valve (PRV) sizing these regulations are related very close to each other. Reference is given to early versions of API, CGA publications [2], [3], [4][1].

The dimensioning of PRV by API, CGA, IGF-Code and IGC-Code [2], [3], [4], [6], [7] are related very close to each other becuase they are based on the same work done in the US in the 50ies [5]. All Codes postulate fire conditions to calculate the generated vapours. In the IGC-Code, IGF-Code and CGA Standard the equivalent air flow (Q) (free air) through the saftey valve is calculated for PRV dimensioning. In API-520 the heat flux into the tank is the criterion for the dimensioning of the PRV. Therfore there is "...no apparent direct comparison between the...formulas" [5].

Nevertheless, it can be shown that these formulas can be converted into each other and that the basic assumption for all formulas is the safe release of the generated vapour in a fire. In addition the assumptions for this fire are identical for API and CGA [5]. Since IGC-Code, IGF-Code are based on API, CGA this is also the case for these Codes. In the following the close relation between these regulations is explained in detail.

---

[1]The current version are dated 2015 for API-520, API-521 and 2022 for CGA-S1.

Chapter 2

Assumed Heat flux

In the following the heat flux definitions for US-Units and SI-Units used by CGA, IGC-Code, IGF-Code and API are explained and the similarities are shown.

2.1   Heat flux according CGA regulations

The total heat flux into an uninsulated tank in BTU/h according CGA is

$$\dot{Q}_{CGA} = \dot{q}_{CGA} \cdot F_{1,CGA} \cdot F_{2,CGA} \cdot A_{CGA}^{0,82} \ [BTU/h] \tag{2.1}$$

The specific heat flux $\dot{q}_{CGA}$ is given by

$$\dot{q}_{CGA} = 34.500,- [BTU/(h \cdot ft^2)] \approx 109 \ [kW/m^2] \tag{2.2}$$

The empirical formula Eq. 2.1 only give correct results if US-Units are used. For a calculation in SI-Units the exponent $0,82$ must be considered to define the specific heat flux. Considering this gives the specific heat flux used by CGA for SI-Units[1]:

$$\dot{q}_{CGA,SI} = 70.961,- [W] \approx 71 \ [kW] \tag{2.3}$$

The unit conversion is detailed in Chapter 6. The final result for the total heat flux in SI-Units ($\dot{Q}_{SI}$) is given by Eq. 6.10.

Using this figure and square meters for the surface area ($A_{CGA,SI}$) leads to the expression:

$$\dot{Q}_{CGA,SI} = 71 \cdot F_{1,CGA} \cdot F_{2,CGA} \cdot A_{CGA,SI}^{0,82} \ [kW] \tag{2.4}$$

2.2   Heat flux according API and comparison to CGA

In the following it is demonstrated that the total heat flux to an uninsulated tank in an area where no prompt fire fighting and drainage of fuel is expected is the same for CGA and API.

Using the formulas given by API it can be expressed by:

---

[1]Note: in this report the European notation for digits is used. "," is used for digits instead of the ".". The "." is used to distinguish thousand values in digits.

$$\dot{Q}_{API} = 21.000, - \cdot \frac{1}{F_{1,API} \cdot F_{2,API}} \cdot A_{API}^{0,82} \ [BTU/h] \tag{2.5}$$

The effective heat flux used in Eq. 2.5 is:

$$\dot{q}_{API} = 21.000, - \ [BTU/(h \cdot ft^2)] \approx 66 \ [kW/m^2] \tag{2.6}$$

For the same reason as given for Eq. 2.2 the value from Eq. 2.6 can only be used if US-Units are used for all parts of formula Eq. 2.5. If SI-Units are used the specific heat flux becomes:

$$\dot{q}_{API,SI} = 43,194 \approx 43 \ [kW/m^2] \tag{2.7}$$

Using SI units for Eq. 2.5 gives:

$$\dot{Q}_{API,SI} = 43 \cdot \frac{1}{F_{1,API} \cdot F_{2,API}} \cdot A_{API,SI}^{0,82} \ [kW] \tag{2.8}$$

The factor 21.000, − "...is not the unit heat flux but is the effective heat flux for a refinery installation. The CGA and API formulas, therefore, become equivalent if one recognizes that the CGA..." and API have "...two invisible F factors"[5].

The F factors in API are exterior environmental factors for refineries considering promot fire fighting and drainage. They become:

$$F_{1,API} \cdot F_{2,API} = 0,60 \tag{2.9}$$

For the transport case covered by CGA no prompt fire fighting and drainage is assumed. Therefore, the F factors are:

$$F_{1,CGA} \cdot F_{2,CGA} = 1,00 \tag{2.10}$$

Considering the API F factors give for Eq. 2.8:

$$\dot{Q}_{API,SI} = \frac{43}{0,60} \cdot A_{API,SI}^{0,82} = 71,7 \cdot A_{API,SI}^{0,82} \tag{2.11}$$

Consequently the specific heat flux in API regulations for SI-Units is:

$$\dot{q}_{API,SI} = 71,7 \ [kW/m^2] \tag{2.12}$$

The above explanation demonstrate that for the same boundary conditions the specific heat flux used by API ($\dot{q}_{API,SI}$, Eq. 2.12) is practically the same as the specific heat flux used by CGA ($\dot{q}_{CGA,SI}$, Eq. 2.3, p. 3).

2.3 Heat flux according IGC-Code and IGF-Code

The IGC-Code and IGF-Code regulations for PRV sizing are based on the CGA regulations. The IGC-Code, IGF-Code use SI-Units.

Considering that the $\dot{q}$ values are rounded figures lead to the conclusion that the API, CGA regulations and IGC-Code, IGF-Code use the same heat flux in SI-Units as given by Eq. 2.3.

$$\dot{q}_{CGA,SI} = \dot{q}_{API,SI} = \dot{q}_{IGC} = \dot{q}_{IGF} = 71 \ [kW/m^2] \tag{2.13}$$

By using the general definition of the specific heat flux ($\dot{q}$) it is possible to calculate directly the mass flow of evaporated vapour which is the basic design value for the PRV sizing.

$$\dot{q} = \dot{m}^* \cdot r_{fg} \;\; ; \left[\frac{kg}{s \cdot m^2} \cdot \frac{kJ}{kg} = \frac{kW}{m^2} \right] \tag{2.14}$$

In Eq. 2.14 the specific mass flow ($\dot{m}$) and the heat of vaporization of the liquefied gas in the tank (r_{fg}) define the specific heat flux ($\dot{q}$).

Chapter 3

Different definitions of surface area in API and CGA

Taking into account the F facors the only real difference between CGA, API regulations is the definition of the heat effected surface are (A). API defines the surface area (A_{API}) as the wetted surface of the tank[1]. CGA defines the surface area (A_{CGA}) as the total surface of the tank.

Both values are the same for a liquid full tank. For a partly filled tank the value required by API can be significabntely smaller than the value required by CGA! The definition of the surface area in the IGC-Code, IGF-Code is identical with the CGA definition (comp. Eq. 3.1).

$$A_{IGC} = A_{IGF} = A_{CGA} \tag{3.1}$$

---

[1]The wetted surface is the surface which is in contact with liquid.

Chapter 4

IGC-Code, IGF-Code requirements

The regulations of Section 8.4 IGC-Code [6] and Section 6.7 IGF-Code [7] are based on the CGA requirements. The requirements of the IGC-Code and the IGF-Code are identical. For this reason the following only refer to the sizing according the IGC-Code.

The fact that the tank is protected when it is installed in a ship is taken into account by using the fire factors F_{IGC}. The fire factors F are the only unique part making IGC-Code, IGF-Code different from CGA regulations.

The heat flux to a ship installed tank (in kW) can be calculated by:

$$\dot{Q}_{IGC} = F_{IGC} \cdot \dot{q}_{CGA,SI} \cdot A_{IGC}^{0,82} \ [kW] \tag{4.1}$$

But this is not the formula given in the IGC-Code Section 8.4. Instead of the heat flux $\dot{Q}$ both - IGC-Code and CGA-S.1 - use the volume flow of "free air" to define the relief valve capacity required[1]. In the following it is demonstrated that Eq. 4.1 and the formulas of the IGC-Code and CGA-S.1 are identical for uninsulated tanks.

For this purpose it is necessary to convert first the heat flux $\dot{Q}$ into a gas flow which is produced by this heat.

$$\dot{Q} = \dot{m}_g \cdot r_{fg} \tag{4.2}$$

The generated mass flow is identical with the IGC-Code, CGA definition when using $F = 1,0$ for an uninsulated, unprotected tank.

$$\dot{Q} = \dot{Q}_{IGC} = \dot{Q}_{CGA,SI} \tag{4.3}$$

$$\dot{m}_g = \frac{\dot{Q}_{IGC}}{r_{fg}} \tag{4.4}$$

This mass flow is the mass flow **of product** which has to be discharged to protect the tank against unacceptable overpressure!

---

[1]"Free Air" is defied as air at standard conditions of $273,15\ K$ and $0,1013\ MPa$.

The requirements in the IGC-Code Section 8.4 are based on the assumptions that the mass flow defined by Eq. 4.4 has to be discharged. The valves are therefore designed to deal with this flow. It should be reminded that the IGC-Code is using SI-Units. For this reason Eq. 2.4, p. 3 must be used to calculate the generated mass flow of vapour ($\dot{m}_g$).

The conversion of the heat flux into the tank to the volume flow of "free air" used in the IGC-Code is hidden in the gas factor (G). The gas factor is defined in the same way by the IGC-Code and CGA. The relation between mass flux ($\dot{m}_g$) and volume flow of free air will be demonstrated in the following.

The generated vapour has to be discharged by the pressure relief valves. For the relief valves the following equation applies[2]:

$$\dot{m}_g = D_g \cdot K_d \cdot K_b \cdot A_0 \cdot p_0 \cdot \sqrt{\frac{Mol_g}{Z_g \cdot T_g}} \tag{4.5}$$

This relation (Eq. 4.5) is valid for the mass flow of air ($\dot{m}_L$) also:

$$\dot{m}_L = D_L \cdot K_d \cdot K_b \cdot A_0 \cdot p_0 \cdot \sqrt{\frac{Mol_L}{Z_L \cdot T_L}} \tag{4.6}$$

In both cases (Eq. 4.5, 4.6) the same vales are assumed. For this reason the formulas can be solved for the valve characteristics A_0, K_d and K_b[3].

$$\dot{m}_g \cdot \frac{1}{D_g \cdot p_0} \cdot \sqrt{\frac{Z_g \cdot T_g}{Mol_g}} = \dot{m}_L \cdot \frac{1}{D_L \cdot p_0} \cdot \sqrt{\frac{Z_L \cdot T_L}{Mol_L}} \tag{4.7}$$

The mass flow of gas ($\dot{m}_g$) is known from Eq. 4.4, p. 7. The only unknown parameter is the mass flow of free air ($\dot{m}_L$). Solving Eq. 4.7 for the mass flow of free air gives:

$$\dot{m}_L = \dot{m}_g \cdot \frac{D_L}{D_g} \cdot \sqrt{\frac{Z_g \cdot T_g \cdot Mol_L}{Z_L \cdot T_L \cdot Mol_g}} \tag{4.8}$$

The missing of the tank pressure in Eq. 4.8 might lead to the misinterpretation that the relation is independent from the tank pressure but the gas properties including tank pressure are included by use of Z_g, T_g, D_g, Mol_g.

To end up with a flow of "free air" the relation between mass- and volume flow is needed. The general relation is:

$$\dot{m} = \dot{V} \cdot \rho \tag{4.9}$$

With Eq. 4.10 the mass flow of free air becomes:

$$\dot{m}_L = \dot{V}_L \cdot \rho_L \ [kg/s] \tag{4.10}$$

In Eq. 4.7 the mass flow of gas can be expressed by using Eq.4.2:

---

[2]comp. also API-520, 4.3.2.1 in 1990 version, or AD-A2 [1]
[3]Minimum orifice area, effective discharge coefficient and capacity correction due to back pressure.

$$\dot{m}_g = \frac{\dot{Q}}{r_{fg}} \tag{4.11}$$

By using the above in Eq. 4.8, p. 8 the volume flow of free air ($\dot{V}_L$) can be expressed by:

$$\dot{V}_L = \frac{\dot{Q}}{r_{fg}} \cdot \frac{1}{\rho_L} \cdot \frac{D_L}{D_g} \cdot \sqrt{\frac{Z_g \cdot T_g \cdot Mol_L}{Z_L \cdot T_L \cdot Mol_g}} \tag{4.12}$$

Still Eq. 4.12 looks very different from any equation of Section 8.4 of the IGC-Code. But in fact, it is identical with the definition of the gas factor (G) in the IGC-Code and in CGA-S.1.

The factor D_g in Eq. 4.12 can be expressed by use of the isentropic exponent (in IGC-Code comp. Sec. 8.4.1.2):

$$D_g = \sqrt{\kappa_g \cdot \left(\frac{2}{\kappa_g + 1}\right)^{(\kappa_g+1)/(\kappa_g-1)}} \tag{4.13}$$

The physical properties of free air are fixed by the regulations. In both regulations the pressure is equal to the atmospheric pressure. Using the standard conditions for air given by the IGC-Code Section 8.4.1.3 of $T_L = 273,15\ K$, $p_L = 0,1013\ MPA$ and $\rho_L = 1,293\ kg/m^3$ and a value for $\kappa_L = 1,402$ to calculate D_L give:

$$\frac{\dot{Q} \cdot D_L}{\rho L} \cdot \sqrt{\frac{Mol_L}{Z_L \cdot T_L}} = \frac{71 \cdot 0,6851}{1,293} \cdot \sqrt{\frac{28,96}{1,0 \cdot 273,15}} = 12,25 \left[\frac{kW \cdot m^3}{kg} \cdot \sqrt{\frac{kg}{kmol \cdot K}}\right] \tag{4.14}$$

Using the factor $12,25 \approx 12,3$ from Eq. 4.14, p. 9 the volume flow $\dot{V}_L$ from Eq. 4.12, p. 9 become:

$$\dot{V}_L = \frac{12,3}{r_{fg} \cdot D_g} \cdot \sqrt{\frac{Z_g \cdot T_g}{Mol_g}} \tag{4.15}$$

By use of $G = \dot{V}_L$, $L = r_{fg}$, $D = D_g$, $Z = Z_g$, $T = T_g$, $M = Mol_g$ and assuming that $12,3 \approx 12,4$ the well known definition of the IGC-Code Section 8.4.1.2, IGF-Code Section 6.7.3.1.1 become visible.

$$G = \frac{12,4}{L \cdot D} \cdot \sqrt{\frac{Z \cdot T}{M}} \tag{4.16}$$

This definition is identical with the definition of the gas factor Gu used by the CGA regulations for uninsulated and insulated containers not meeting the requirements of CGA-S.1.[4]

In IGC-Code, IGF-Code and CGA regulations it is stipulated that the maximum pressure will be 120 % of the pressure at which the pressure relief valve (PRV) is set. The assumed temperature of the gas used by Eq. 4.16, p. 9 is the boiling temperature corresponding to this pressure[5]. The isentropic exponent for calculation of D is usually taken at standard atmospheric pressure and temperature. The physically

---

[4] comp. Section 5.3.4, CGA-S.1.2
[5] Not the boiling temperature at Maximum Allowable Relief Valve Setting (MARVS)!

more correct value is the isentropic exponent at saturation line for T at $1, 2 \cdot MARVS$. The error by using the atmospheric conditions is to the safe side[6].

The gas factor G is finally used to define the volume flow of "'free air" for the IGC-Code, IGF-Code by

$$Q = F \cdot G \cdot A^{0,82} \qquad (4.17)$$

and for an uninsulated tank according CGA-S.1, Section 5.3.2 with:

$$Q = Gu \cdot A^{0,82} \qquad (4.18)$$

As the F factor for uninsulated tanks on deck is $F = 1, 0$ in IGC-Code, IGF-Code the equations Eq. 4.17 and Eq. 4.18 are equivalent for this tank arrangement. The Fire factor F in the IGC-Code, IGF-Code is used to consider the different level of protection against a fire. E.g., for an insulated tank in a tank hold the factor is $F = 0, 2$.

---

[6]Higher flow.

Chapter 5

Summary

The most common procedures for evaluating the required capacity of pressure relief valves for tanks with liquefied gases in case of a fire are API-520/521 [2], [3] and CGA-S.1 Part 1., 2, 3. [4]. For ships carrying liquefied gases in bulk Section 8.4 of the IGC-Code [6] is mandatory. For ships using Liquefied Natural Gas (LNG) as fuel the IGF-Code Section 6.7.3 [7] is mandatory. The related rules in IGC-Code and IGF-Code are identical.

The foundation of these regulations have been developed during the 1950*ies* by API and CGA in the United States [5]. The heat fluxes assumed are amximum values from fire tests which were available at that time.

The maximum value assumed by CGA and API for the heat input through the wetted surface of a tank has been compiled from the fire tests with $34.000, - [BTU/(h \cdot ft^2)]$ which is equal to $109, - [kW/m^2]$[1]. These values comply with the maximum heat flux from a hydrocarbon fire and therefore can be used for the evaluation of hydrocarbon fires in general.

The API regulations consider prompt fire fighting and drainage for their applications in refineries. Therefore, the value for the heat flux used by API is reduced by the factor $0, 6$ to $21.000, - [BTU/(h \cdot fr^2)]$ which is equal to $66, - [kW/m^2]$[2].

For using the above heat fluxes for tank safety valve sizing API and CGA consider that in a practical cases a tank is not completely engulfed by a fire. For this reason the effected surface is reduced by the exponent $0, 82$[3]. For the calculation in SI-Units as it is done in the IGC-Code and IGF-Code this exponent has to be considered. For this reason the specific heat flux used in these Codes is $71 \ [kW/m^2]$ and **NOT** $109 \ [kW/m^2]$ to end up with the same results[4].

IGC-Code, IGF-Code and CGA regulations are using the total tank surface as heat exchanging surface. API is using the part of the tank surface which is wetted by liquid. API does not consider the tank surface in the gas phase of the tank. For a liquid full tank the results of all regulations are the same. For a tank filled to the maximum allowable filling level the results are also nearly the same. A partly

---

[1]comp. Eq. 2.2, p.3

[2]comp. Eq. 2.6, p. 4

[3]comp. Eq. 2.1, p. 3, Eq. 2.5, p. 4

[4]comp. Eq. 2.3, p. 3. For refinery installations comp. Eq. 2.7, p. 4

filled tank[5] may have significantly lower valve capacities if the valves are designed according API regulations.

API is calculating directly the mass of liquefied gas evaporated by the fire and is using this figure directly to define the required minimum orifice area of the relief valve (PRV). CGA, IGC-Code and IGF-Code calculate an equivalent flow of air at standard atmospheric conditions[6]. The procedure to calculate the orifice area is not part of CGA, IGC-Code and IGF-Code regulations. Valve manufacturers give this volume flow of "free air" in their valve specifications.

In practice the use of a volume flow is often confusing for designers and even approval engineers. The way from generated vapour to volume flow of "'free air" is outlined in this paper.

---

[5]E.g., a tank with low maximum allowable loading limit.

[6]Note that "standard atmospheric conditions" always use the same pressure of $1,013\ bar$ but may be different with regard to the used air temperature. Common values are $0\ C^o$, $15\ C^o$ and $25\ C^o$. The consequence are different air densities.

Chapter 6

Add on: From US to SI-Units

The basic formula used by Cummings (comp. [5]) and therefore in the related API, CGA, IGC-Code and IGF-Code regulations can be written as:

$$\dot{Q}_{US} = \frac{Q}{H \cdot A} = \dot{q}_{US} \cdot A_{US}^{0,82} \ [BTU/h] \tag{6.1}$$

In Eq. 6.1 $A = A_{US}$ is the wetted surface[1] in $[ft^2]$, $\dot{Q}_{US}$ is the heat flux in $[BTU/h]$. Cummings defines A_{US} as a factor without a unit[2]. Therfore $H \cdot A$ is needed to match the unit equation. $\dot{q}_{US}$ is named "fuel factor" with a value of $34.000, - [BTU/(h \cdot ft^2)]$ [5], p. 136, 137).

To define the conversion factor to SI-Units Eq. 6.1 is written in the form:

$$\dot{Q}_{US} = a_1 \cdot \dot{q}_{SI} \cdot (a_2 \cdot A_{SI})^{0,82} \ [BTU/h] \tag{6.2}$$

which is equal to:

$$\dot{Q}_{US} = a_1 \cdot \dot{q}_{SI} \cdot a_2^{0,82} \cdot A_{IS}^{0,82} \ [BTU/h] \tag{6.3}$$

For Eq. 6.2, 6.3 the specific heat flux $\dot{q}_{SI}$ is given in W and the wetted surface A_{SI} in m^2. The total heat flux can be converted to SI units by:

$$\dot{Q}_{SI} = a_3 \cdot \dot{Q}_{US} \ [W] \tag{6.4}$$

Combining Eq. 6.3, 6.4 results in:

$$\dot{Q}_{SI} = a_3 \cdot a_1 \cdot a_2^{0,82} \cdot \dot{q}_{SI} \cdot A^{0,82} \ [W] \tag{6.5}$$

The conversion factors are:

factor	value	conv. from	conv. to
a_1	$1/3, 1546 = 0, 316997$	$W/m^2 \rightarrow$	$BTU/(ft^2 \cdot h)$
a_2	$1/0, 092903 = 10, 763915$	$m^2 \rightarrow$	ft^2
a_3	$0, 29308$	$BTU/h \rightarrow$	W

Table 6.1: Conversion factors

Using the values from Tab. 6.1 the final conversion factor becomes:

---

[1]Wetted by liquid in the tank.
[2]comp. p. 136 in [5]

$$a_4 = a_1 \cdot a_2^{0,82} \cdot a_3 = 0,652015 \tag{6.6}$$

From this Eq. 6.4, p.13 becomes:

$$\dot{Q}_{SI} = a_4 \cdot \dot{q}_{SI} \cdot A_{SI}^{0,82} \ [W] \tag{6.7}$$

The specific heat flux is named "fuel factor". The value is:

$$\dot{q}_{SI} = 3,1546 \cdot q_{US} = 108.833,7 \ [W/m^2] = 109 \ [kW/m^2] \tag{6.8}$$

The fuel factor is a constant value and therefore can be incorporated into the unit conversion. The finally used specific heat flux in SI-Units is:

$$\dot{q}_{SI}^* = a_4 \cdot \dot{q}_{SI} = 70.961,2 \ [W/m^2] = 71 \ [kW/m^2] \tag{6.9}$$

The final basic formula in $SI - Units$ for a bare tank in a hydrocarbon pool fire is:

$$\dot{Q}_{SI} = \dot{q}_{SI}^* \cdot A_{SI}^{0,82} [kW] \tag{6.10}$$

It should be noted that the value of 71 $[kW/m^2]$ (Eq. 6.9) include the assumption that the tank is not completely engulfed by the fire. For small tanks or pipes which are completely engulfed by a fire 109 $[kW/m^2]$ (Eq. 6.8) is the correct value.

Acronyms

API American Petroleum Institute

CGA Compressed Gas Association

IGC-Code International Code for the Construction and Equipment of Ships Carrying Liquefied Gases in Bulk

IGF-Code International Code of Safety for Ships using Gases or other Low-Flashpoint Fuels

LNG Liquefied Natural Gas

MARVS Maximum Allowable Relief Valve Setting

PRV Pressure Relief Valve

Bibliography

[1] *AD-Merkblatt A2: Sicherheitseinrichtungen gegen Drucküberschreitung - Sicherheitsventile.* Richtlinie. Beuth Verlag GmbH, Berlin, 1990.

[2] *API RP 520 Part 1 and 2; Sizing, Selection and Installation of Pressure-Relieving Devices in Refineries; 5th edition.* API Guideline. The American Petroleum Institute, Washington D.C., 1990.

[3] *API RP 521; Guide for Pressure-Relieving and Depressuring Systems; 3th edition.* API Guideline. The American Petroleum Institute, Washington D.C., 1990.

[4] *CGA S-1-1989, 1980, 1980; Pressure Relief Device Standards; Part-1-Cylinders for Compressed Gas; Part 2-Cargo and Portable Tanks for Compressed Gases; Part 3-Compressed Gas Storage Containers.* Standard. Compressed Gas Association, CGA; Chantilly, Virginia, US.

[5] Frank J. Heller. "Safety Relief Valve Sizing: API Versus CGA Requirements Plus A New Concept For Tank Cars". In: *API-Refinery Department, 1983-Proceedings* 6.2 (), pp. 123–135.

[6] *IMO IGC-Code, International Code for the Construction and Equipment of Ships Carrying Liquefied Gases in Bulk.* 2016th ed. IMO London.

[7] *IMO IGC-Code, International Code of Safety for Ships using Gases or other Low-Flashpoint Fuels.* 2016th ed. IMO London.

Glossary

Air Pollution is used here for emissions to air which have a short term negative effect to human health and ambient. Mainly: SO_2, NO_x, PM. (Sect. 1.4, p. 5).

Cargo Area The cargo area on gas carriers includes all spaces with cargo tanks and cargo related process equipment. This is the main part of the ship excluding the deck house and the bow area where the anchors and bow mooring winches are installed. (Sect. 3.2, p. 19).

Deflagration A deflagration is a combustion event with a flame velocity below the sound velocity. A deflagration is related to a pressure increase in the *mbar* range. In open space nearly all flammable gas/air mixtures may create a deflagration but no detonation. (Sect. 4.3, p. 26).

Detonation A detonation is a combustion event with a flame velocity equal to or above the sound velocity. A deflagration creates pressure increases of 8 *bar* and more above the pressure in the space if the space was at ambient pressure before. To the knowledge of the author Hydrogen/air mixtures combined with a jet release of Hydrogen in an ambient with obstacles which increase turbulence is the only flammable gas/air mixture which can create explosions in an open space. (Sect. 4.3, p. 26).

Explosion An Explosion is generally understood as an violent combustion event. In most cases the term summarize deflagration and detonation behaviour. (Sect. 4.3, p. 26).

GHG emissions is used here for emissions to air which have a notable Green House Gas effect to the atmosphere. Mainly: CO_2, CH_4 and N_2O. (Sect. 1.5, p. 7).

Fatal Crack A fatal crack is a crack in a structure which may propagate without additional load cycles. The criterion for a fatal crack is that it has a length longer than the critical crack length. For a shorter crack the propagation stops and needs a load cycle to propagate. Beyond the critical crack length no load cycle is needed for crack propagation. (Sect. 6.2, p. 40).

© The Editor(s) (if applicable) and The Author(s), under exclusive license to Springer
Nature Switzerland AG 2025
G. Würsig, *The Safety Principles for the Use of Low Flashpoint Fuels in Shipping*,
Synthesis Lectures on Ocean Systems Engineering,
https://doi.org/10.1007/978-3-031-64174-9

Fatigue Crack A fatigue crack is a crack caused by material fatigue. The pressure, weight and temperature load cycles of a tank have a large influence on the fatigue. (Sect. 6.2, p. 40).

FL The maximum Filling Limit (FL) according IGF-Code, IGC-Code is the maximum value of liquid allowed at any time to be in the tank. It is a relative value related to the total geometrical volume of the tank. The FL considers the liquid expansion related to temperature changes in the liquid. Comp. also maximum Loading Limit (LL). The FL is a regulation requirement and not a design value. (Sect. 6.6.2, p. 53).

Flashpoint is defined by the temperature at which a flammable gas air mixture above a burnable liquid can be ignited by a defined ignition source. (p. vii).
"Flashpoint is the temperature in degrees Celsius (closed cup test) at which a product will give off enough flammable vapour to be ignited, as determined by an approved flashpoint apparatus." SOLAS: Ch II-2, [1].

FMEA is a systematic approach to identify failures, their conditions to happen and the consequences they might have. The FMEA is done at a late design stage when the system details are known. In the best case, only a last round for modification is needed to include the FMEA results into the design. In the very best case, the FMEA and its documentation can be used as a tool to maintain the system, improve the design and the safety. (Sect. 2.2, p. 13).
To be honest this is a wish of the author in order to improve safety and to save costs. It is not common in shipping. Most time the involved parties are happy to demonstrate (e.g. to the Classification Society) that the system is "ok" at the time when the FMEA was done.

HAZID is a screening method to identify risks in an early stage of a system design process. It is often done in moderated workshops involving the different parties like designer, end user, operator. (Sect. 2.2, p. 13).

IMO-CCC IMO Sub-Committee on Carriage of Cargoes and Containers. CCC reports to the IMO MSC (Maritime Safety Committee) which is the legal decision body for IMO legislation on ship safety. (Sect. 1.3, p. 4). Note: the meeting number of IMO bodies is added at the end of the shortcut.

IMO Interim Guideline voluntary IMO Guideline intended to be used as design guidance for subjects not covered by the international law which is the SOLAS Convention. (Sect. 1.2, p. 3).

LL The maximum loading Limit LL according IGC-Code, IGF-Code is the maximum liquid content allowed to be loaded into a tank related to the geometrical volume of the tank. The LL is a design value considering the maximum FL, tank pressure, liquid temperature. (Sect. 12.1.2, p. 101).

Moss type tank The Moss company was the first company which designed a type B tank. Moss tanks have a spherical shape which makes the required calculations and test procedures relatively easy. In fact, the type B tank requirements have been developed based on the Moss design development. (Sect. 6.1, p. 35).

Power to X Power to X is a synonym for energy carriers produced from Hydrogen (in most cases from renewable or nuclear energy) and one or more other molecules. The most relevant PtX energy carriers are Hydrogen itself (H_2), Methane (CH_4), Methanol (CH_3OH), Fischer Tropsch distillates with a high number of carbon molecules which can be regarded as Diesel like energy carriers (C_iH_n) and Ammonia (NH_3). (Sect. 1.5, p. 7).

Risk is defined as a relation between the likelihood of an event and the consequences this event will have if it occurs. (Sec. 2, p. 9)

SPB design The SPB design has been developed in Japan by IHI since the 1960s for LPG carriers. In the early 1980s IHI stated the development of SPB tanks for LNG carriers. The first LNG carriers with SPB tanks went into operation in 1993 (87.500, $-$ m^3 SPB LNG carriers Polar Eagle and Arctic Sun). (Sect. 6.3, p. 43).

Reference

1. SOLAS 2020 Consolidated Edition; ISBN: 978-92-801-1690-8, IMO, London